industrial mbrs

membrane bioreactors for industrial wastewater treatment

Simon Judd
with Claire Judd

Professor Simon Judd lectures at Cranfield University in the UK, where he is Professor of Membrane Technology, and Qatar University in the Middle East, where he holds the post of the Maersk Oil Professorial Chair in Environmental Engineering.

Claire Judd is manager of The MBR Site and a professional copy editor.

Judd and Judd Ltd
Registered office:
1a The Avenue, Flitwick, Bedfordshire MK45 1BP, United Kingdom

Judd and Judd Ltd is a company registered in England and Wales,
registered number 8082403

Co-published by IWA Publishing
Alliance House, 12 Caxton Street, London SW1H 0QS, UK
Tel: +44 (0)20 7654 5500, Fax: +44 (0)20 7654 5555
publications@iwap.co.uk
www.iwapublishing.com
ISBN13: 9781780407036

ISBN 9-781780-407036

www.thembrsite.com

Preface

An increasing number of books exist based on membrane bioreactor (MBR) technology. Apart from the Butterworth-Heinemann/Elsevier reference texts *The MBR Book* first and second editions (respectively 2006 and 2011), there are also at least two books published by WEFpress (*Membrane Systems for Wastewater Treatment*, 2006, and *Membrane BioReactors WEF Manual of Practice* No. 36, 2011), and a number from IWA Publishing (*Membrane Bioreactors: Operation and Results of an Mbr Wastewater Treatment Plant* edited by Bentem, Petri and Schyns, 2007; Brepol's *Operating Large Scale Membrane Bioreactors for Municipal Wastewater Treatment* from 2010, and the recently published *Membrane Biological Reactors* edited by Hai, Yamamoto and Lee, 2014). The subject of MBRs is included in books on both wastewater treatment/reuse and membrane technology (too numerous to mention, but a recent example encompassing both is Wachinski's *Membrane Processes for Water Reuse*, McGraw-Hill, 2012).

It is, however, notable that many of these books are focused predominantly on municipal rather than industrial effluents. Whilst the MBR technology itself is independent of the application, key design parameters and pre-treatment requirements can differ appreciably and industrial effluents *per se* pose their own challenges. The quality of the effluent varies significantly between sectors, within sectors, and even temporally, diurnally or seasonally for a specific installation. As such the application of MBRs to industrial effluents merits special attention.

As with our previous books we have tried to maintain a practical focus throughout. Indeed, there is no section focused on research and development at all in this book: those seeking such a perspective are directed to Hai et al.'s recent book. Instead, there is a general introduction (Chapter 1), a summary of MBR technology and the key design parameters and cost equations (Chapter 2), a review of industrial effluents (Chapter 3), a compilation of commercially available MBR technologies (Chapter 4) and over fifty case studies (Chapter 5). Processed data from the case studies are presented in the final section of the book.

The terminology and abbreviations are defined in the Abbreviations (x) and Symbols (xi). Units used are exclusively SI: a table of conversion factors for US units is given in Annex 2.

During the course of our research for *Industrial MBRs*, we have been in contact with over 100 contributors from over 60 companies worldwide (page iv). It is therefore almost inevitable that there will be a certain number of (hopefully small) errors and omissions in the text. While we have gone to every effort to try to ensure the accuracy and completeness of the content of this book, please note we cannot be held liable for any errors or omissions. Please notify us at info@thembrsite.com of any issues so we can correct the situation for future editions.

Finally, we would like to thank the many contributors to this book, listed in the Contributors section (page iv), as well as our generous sponsors (page iii). Without their support, both this book and The MBR Site would not be possible.

Simon and Claire Judd

About the author

Professor Simon Judd lectures at Cranfield University in the UK and Qatar University in the Middle East. He has over 22 years' experience in teaching the fundamentals of water and wastewater technologies, and has spent almost 20 years working in membrane bioreactor technology in a teaching, research and consultative capacity.

Industrial MBRs is Simon's seventh book on the subject of water and wastewater treatment. His most recent publications are *The MBR Book: Principles and Applications of Membrane Bioreactors for Water and Wastewater Treatment* (second edition), published by Elsevier in 2011, and *Watermaths* (second edition), a textbook for undergraduates and practitioners on maths for water and wastewater treatment technologies, published by Judd and Judd Ltd in 2013.

Simon continues to conduct research into municipal water and wastewater treatment generally, also specialising in produced water treatment technologies for the oil and gas sector, and provides consultancy and training on water and wastewater treatment to clients across the globe.

The MBR Site

The MBR Site (thembrsite.com) is a specialist website for those interested in membrane bioreactors for water and wastewater treatment – practitioners, researchers and technology providers.

The website focuses on practical information and guidance, including directory lists of membrane products, technology suppliers, consultants and contractors, and reference installations, as well as Simon's MBR blog and spotlight features on membrane bioreactor technology in practice.

The MBR Site is managed by Claire Judd, co-editor of this book.

The MBR Group

The MBR Group on LinkedIn.com is a lively discussion forum of approaching 5,000 professionals worldwide (as at September 2014), managed by Simon and Claire Judd.

Sponsors

We would like to thank the following companies (listed here in alphabetical order) for their generous sponsorship of *Industrial MBRs* and for their continuing support of The MBR Site:

Aquabio Ltd/Freudenberg Filtration Technologies	aquabio.co.uk
Ecologix Technologies Asia Pacific, Inc.	ecologix.com
Econity Co., Ltd.	econity.com
Kubota Membrane Europe Ltd	kubota-mbr.com
Memstar Pte Ltd/United Envirotech Ltd	memstar.com.sg
Ovivo/GLV Group	ovivowater.us
Pentair Inc.	pentair.com
Shanghai SINAP Membrane Tech Co.,Ltd.	sh-sinap.com
Toray Membrane Europe AG	toraywater.com

In addition, we would also like to thank the following companies for their generous support of The MBR Site:

A3 Water Solutions GmbH	a3-gmbh.com
ADI Systems Inc.	adisystemsinc.com
Berghof Membrane Technology GmbH & Co. KG	berghof.com
GE Water & Process Technologies	gewater.com
Grundfos BioBooster A/S	grundfos.com
MICRODYN-NADIR GmbH	microdyn-nadir.com
Shanghai MegaVision Membrane Engineering & Technology Co.,Ltd	china-membrane.com
Veolia-Biothane	biothane.com
WEHRLE Umwelt GmbH	wehrle-umwelt.com

Contributors

The authors would like to thank the following contributors to *Industrial MBRs*, all of whom have been extremely generous in their time and support for this project. We recognise that there have been a great number of people involved in providing details of industrial sectors, technologies, applications and installations, and that we may have inadvertently omitted to list a number of contributors who also deserve to be included here. If this is the case, we apologise profusely for our oversight, as our intention is not to offend.

A special thanks to Dan Golea at Cranfield University for assisting with the acquisition of data.

Name	*Affiliation*	*Section(s) contributed to*	
Miguel Aljovin de Losada	Esmeralda Corp SAC	5.1.14	
Sebastian Andreassen	LiqTech International A/S	4.1.11	
Roy Arviv	GE Water & Process Technologies	5.2.9	
Galina Atanasova	Evoqua Water Technologies, LLC	4.2.3	
Junichi Baba	Toray Industries, Inc.	5.1.14	
Jessica Bengtsson	Alfa Laval	4.1.1	
Zhuo Bang Cai	Sinopec Guangzhou Company	5.2.1	
Lutz Bungeroth	ItN Nanovation AG	4.1.7	
Chia-Yuan Chang	Chia Nan University	5.3.6	
Wei Chen	Wartsila Water Systems Ltd.	3.8	5.7.2
Scott Christian	ADI Systems Inc.	5.1.18	
Michael Chua	Ceraflo Pte Ltd.	4.1.4	
Jason Diamond	GE Water & Process Technologies	5.1.1	5.1.7
Miroslav Dohnal	Zena s.r.o.	4.2.18	
Antonio Ferreira	Valorsabio	5.3.4	
Maria Luisa Flores	Frigorifico JOSAC	5.1.14	
Beth Galic	QUA Group	4.1.19	
Don Gilhooly	Canada Malting	5.1.1	
Jordi Gomà-Camps Travé	Gomà-Camps S.A.U.	5.4.3	
Steve Goodwin	Aquabio Ltd, Freudenberg Filtration Technologies	4.3.3.1	
Ben Gould	Alfa Laval Copenhagen A/S	4.1.1	
Shannon Grant	ADI Systems Inc.	5.1.18	
Péter Groszmann	Toray Membrane Europe AG	4.1.22	5.1.14
Squall Gu	Benenv Co., Ltd	4.1.2	
Torsten Hackner	Huber SE	4.1.6	
Ge Hailin	Memstar Pte Ltd.	4.2.10	
Minoru Hayashi	Microza & Water Processing Division, Asahi Kasei Chemicals Corporation	5.1.11	
Justin He	Litree Purifying Technology Co., Ltd.	4.2.8	
Barry Heffernan	Biothane Systems International, Veolia Water Solutions & Technologies	5.1.17	
Matsumoto Hideki	Asahi Kasei Chemicals Corporation	4.2.1	
Caroline Hoffmann	MICRODYN-NADIR GmbH	4.1.16	
Gerin James	Evoqua Water Technologies, LLC	4.2.3	
Bjarne Jensen	KMC	5.1.8	
Cao Guo Jia	Taixing Binjiang Wastewater Treatment Co., Ltd.	5.8.2	
Oriol Jimenez	Berghof Membrane Technology GmbH & Co. KG	5.1.9	5.5.4
Thérèse Johansson	ÅF AB	3.5	
Emmanuel Joncquez	Alfa Laval	4.1.1	5.1.8
Gunnar Junker	Supratec Filtration GmbH & Co. KG	4.1.21	
Mito Kanai	Kubota Membrane Europe Ltd	4.1.9	
Stephen Katz	GE Water & Process Technologies	4.2.4	
Gyuwan Kim	Econity Co., Ltd.	4.2.2	
John Kirkpatrick	Basic American Foods	5.1.7	
Udo Kolbe	Ultura Water	4.3.1.5	
Robert Körner	WEHRLE Umwelt GmbH	4.3.3.4	
Henning Krajinski	Ultura Water	4.3.1.5	
Christina Kuhn	Toray Membrane Europe AG	4.1.22	
Christoph Kullmann	Xylem Water Solutions	4.3.1.3	
Peter Küppers	Toray Membrane Europe AG	4.1.22	
Annabell Lanz	Aquabio Ltd, Freudenberg Filtration Technologies	4.3.3.1	
Asun Larrea	Praxair Inc.	5.3.5	
Ahreum Lee	Kolon Industries, Inc.	4.2.7	

Adams Li	Shanghai SINAP Membrane Tech Co., Ltd.	4.1.20	5.2.6	5.6.1
Fangyue Li	ROCHEM	5.7.3		
Guoming Liang	Shanghai SINAP Membrane Tech Co., Ltd.	5.6.1		
Lily Lien	Memstar Pte Ltd.	4.2.10		
Justin Lin	Dept. Environmental Engineering and Science, Feng Chia University	5.2.2		
Dennis Livingston	Ovivo	5.3.2	5.3.3	
David Lo	Ecologix Technologies Asia Pacific, Inc.	4.1.5		
Kevin Lo	Ecologix Technologies Asia Pacific, Inc.	4.1.5	5.1.12	
Carey Loy	Hyflux Ltd.	4.2.5		
Michael Lyko	MICRODYN-NADIR GmbH	4.1.16		
Maggie Ma	Tianjin Motimo Membrane Technology Co., Ltd.	5.2.4	5.8.3	
Kimberly Mathis	Formerly of Ovivo	5.3.2	5.3.3	
Elena Meabe	Likuid Nanotek	5.1.15		
Mr. Meyer	WEHRLE Umwelt GmbH	5.6.4		
Victor Monsalvo	Autonomous University of Madrid	5.3.1		
Kevin Moore	Vietnam Waste Solutions	5.6.3		
Toshi Motohori	Kubota Membrane Europe Ltd	4.1.9		
V J Nathan	QUA Group	4.1.19		
Søren Nøhr Bak	Grundfos BioBooster A/S	5.1.6		
Katsumi Ooshita	Sumitomo Electric Fine Polymer, Inc.	4.2.16		
Björn Otto	Microclear, newterra GmbH	4.1.17		
Jennifer Pawloski	GE Water & Process Technologies	4.2.4		
Darren Reed	Xylem Inc.	4.3.1.3		
Steffen Richter	A3 Water Solutions GmbH	4.1.13		
Mark Robinson	WEHRLE Umwelt GmbH	4.3.3.4	5.6.3	5.6.4
Christian Roloff	Martin Membrane Systems AG	4.1.12		
Sean Roop	Industrial and Engineering Consultants, LLC	3.3		
Archie Ross	Dynatec Systems Inc.	4.3.3.3	5.6.5	
Werner Ruppricht	MICRODYN-NADIR GmbH	5.1.16		
Monica Santos	Amgen Inc.	5.3.2		
Kathrin Sauter	MICRODYN-NADIR GmbH	4.1.16	5.1.16	
Dirk Schlemper	Koch Membrane Systems Inc.	4.2.6		
Wolfgang Schoene	MICRODYN-NADIR GmbH	5.1.16		
Antonio Sempere	Likuid Nanotek	5.1.15	5.3.5	
Claudia Silvius	Berghof Membrane Technology GmbH & Co. KG	5.1.9	5.5.4	
Marion Slaghuis	Pentair Ltd.	4.3.1.4		
Connie Smith	ADI Systems Inc.	5.1.18		
Jens Sonntag	MaxFlow Membran Filtration GmbH, A3 Water Solutions GmbH	5.1.10		
Karin Sutherland	Thyssen Krupp Marine Systems	5.7.1		
Chindia Tang	Hangzhou Microna Membrane Technology Co., Ltd.	4.2.9		
Nancy Tao	Shanghai MegaVision Membrane Engineering & Technology Co., Ltd.	4.1.14	5.3.4	
Liu Tao	Sinopec Luoyang Company	5.2.7		
Rick te Lintelo	Berghof Membrane Technology GmbH & Co. KG	5.1.9	5.5.4	
Yuri Tsutomu	Mitsubishi Rayon Co., Ltd.	4.2.11		
Alexander Tsypin	Intel	5.2.8		
Farid Turan	MBR Solutions Limited	3.2		
Sjaan van der Pol	Berghof Membrane Technology GmbH & Co. KG	4.3.1.1	4.3.3.2	
Ronald van't Oever	Pentair Ltd.	4.3.1.4	5.1.2-5	
Bart Verrecht	Hach-Lange	2.7.2.1		
Daniella Weisbort Andreassen	Grundfos BioBooster A/S	4.3.2.1		
Simon Yong	NOVO Envirotech (Guangzhou) Ltd.	5.2.1	5.2.7	
Low Weiyou	Ceraflo Pte Ltd	4.1.4		
Steve Wilcox	Aquabio Ltd, Freudenberg Filtration Technologies	4.3.3.1	5.1.13	
Eric Wildeboer	Berghof Membrane Technology GmbH & Co. KG	4.3.1.1	4.3.3.2	
Jeff Williams	Shepherd Neame Brewery	5.1.5		
Chris Woods	Aquabio Ltd, Freudenberg Filtration Technologies	5.1.5		
Dongxu Yan	Layne Christensen	4.2.16	5.4.2	
Hang Yang	Mono Pumps	3.8		
Yiming Zeng	Superstring MBR Technology Corp.	4.2.17		
Amy Zhang	Tianjin Motimo Membrane Technology Co., Ltd.	4.2.14		
Nancy Zheng	Tianjin Motimo Membrane Technology Co., Ltd.	5.2.4		
István Zsirai	GE Water & Process Technologies	5.2.8		

CONTENTS

ABBREVIATIONS

AD	anaerobic digestion
ADF	average daily flow
ADUF	anaerobic digestion-ultrafiltration
AE or Ae	aerobic
AF	anaerobic filter
A-L sMBR	air-lift sidestream MBR
AN or An	anaerobic
An iMBR	anaerobic immersed MBR
An sMBR	anaerobic sidestream MBR
AO or Ao	anoxic
API	active pharmaceutical ingredient
API	American Petroleum Institute
ASP	activated sludge process
BAF	biological aerated filter
BAT	best available technology
BNR	biological nutrient removal
BOD	5-day biochemical oxygen demand
BPR	biological phosphorus removal
CA	cellulose acetate
CAGR	compound annual growth rate
CAPEX	capital expenditure
CAS	conventional activated sludge
CBD	coarse bubble diffuser
CEB	chemically enhanced backflush
CFU	colony-forming units
CFV	crossflow velocity
CIP	clean in place
COD	chemical oxygen demand
CSTR	continuous stirred tank reactor
CTMP	chemical thermomechanical process
DAF	dissolved air flotation
DO	dissolved oxygen
DOC	dissolved organic carbon
DS	dry solids
EPC	engineering, procurement and construction
EPS	extracellular polymeric substances
EQ	equalisation
FBD(A)	fine bubble diffuser (aeration)
Flocs	flocculated particles
F:M	food to micro-organism (ratio)
FOG	fats, oils and grease
FS	flat sheet
GFD	gallons per square foot per day
HDPE	gigh-density polyethylene
HF	gollow fibre
HRT	gydraulic retention time
Hz	Hazen
IAF	induced air flotation
IFAS	integrated fixed film activated sludge
iFS	immersed flat sheet
IGF	induced gas flotation
iHF	immersed hollow fibre
iMBR	immersed membrane bioreactor
LMH	litres per m^2 per hour
MBBR	moving bed bioreactor
MBR	membrane bioreactor
MC	multi-channel
MF	microfiltration
MGD	megagallons per day
MLD	megalitres per day
MLE	modified Ludzack-Ettinger
MLSS	mixed liquor suspended solids
MLVSS	mixed liquor volatile suspended solids
MT	multi-tube
N	nitrogen
NF	nanofiltration
NPV	net present value
O&G	oil and gas
O&M	operation and maintenance
OEM	original equipment manufacturer
OLR	organic loading rate
OPEX	operating expenditure
OTE	oxygen transfer efficiency
OUR	oxygen utilisation rate
P	phosphorus
PA	polyamide
PAN	polyacrylonitrile
PCP	personal care products
PDF	peak daily flow
PE	polyethylene
PES	polyethylsulphone
PP	polypropylene
PS	polysulphone
psi	pounds force per square inch
PTFE	polytetrafluoroethylene
PVA	polyvinyl alcohol
PVDF	polyvinylidene difluoride
PW	produced water
RAS	return activated sludge
RBC	rotating biological contactor
RHS	right-hand side (of equation)
RO	reverse osmosis
SAD	specific aeration demand
SAF	submerged aerated filter
SBR	sequencing batch reactor
SCFM	standard cubic feet per minute
SDI	silt density index
SED	specific energy demand
sMBR	sidestream membrane bioreactor
SMP	soluble microbial product
SOTE	standard oxygen transfer efficiency
SRT	solids retention time
TDS	total dissolved solids
TF	trickling filter
TIPS	thermal-induced phase separation
TKN	total Kjeldahl nitrogen
TMP	thermomechanical process
TMP	transmembrane pressure
TN	total nitrogen
TOC	total organic carbon
TOTEX	total expenditure
TSS	total suspended solids
UASB	upflow anaerobic sludge blanket
UF	ultrafiltration
USCG	US Coast Guard
USEPA	US Environmental Protection Agency
UV	ultraviolet
VOC	volatile organic carbon
VSD	variable speed drive
VSS	volatile suspended solids
WAS	waste activated sludge
ZLD	zero liquid discharge

SYMBOLS

C'_A	oxygen concentration, mg/L
D_{O2}	oxygen demand, mg/L
dP/dt	fouling rate, pressure per unit time
E	specific energy demand per unit volume permeate, kWh/m^3
E_A	specific energy demand for aeration per unit volume permeate, kWh/Nm^3
$E_{A,m}$	specific energy demand for membrane air scouring per unit volume permeate, kWh/m^3
E'_A	specific energy demand for aeration per unit volume air, kWh/Nm^3
$E'_{A,bio}$	specific energy demand for biological aeration per unit volume air, kWh/Nm^3
$E'_{A,m}$	specific energy demand for membrane air scouring per unit volume air, kWh/Nm^3
E_{contr}	specific energy demand for process control per unit volume permeate, kWh/m^3
E_L	specific energy demand for liquid/sludge pumping per unit volume permeate, kWh/m^3
$E_{L,bio}$	specific energy demand for biological aeration per unit volume permeate, kWh/m^3
$E_{L,chem}$	specific energy demand for chemical cleaning per unit volume permeate, kWh/m^3
$E_{L,m}$	specific energy demand for permeation per unit volume permeate, kWh/m^3
$E_{L,sludge}$	specific energy demand for sludge pumping per unit volume permeate, kWh/m^3
E_{mix}	specific energy demand for mixing per unit volume permeate, kWh/m^3
E_{other}	specific energy demand for other operations, kWh/m^3
F_A	specific area footprint, m^3/h per m^2 area (i.e. m/h)
$F:M$	food:micro-organism ratio
F_V	specific volume footprint, m^3/h per m^3 volume (i.e. h^{-1})
i	discount rate (rate of return on the investment were the capital sum to be invested)
J	flux, L/(m^2.h)
J_b	backflush flux, L/(m^2.h)
J_{net}	net flux, L/(m^2.h)
k	constant in Equation 10
L	specific cost per m^3 treated water
L_C	specific cost per m^3 treated water for chemicals consumption
L_E	specific cost of electrical energy per kWh
L_L	specific cost per m^3 treated water for labour
L_M	specific cost per m^2 membrane area
L_W	specific cost per m^3 treated water for waste disposal
$O_{2,COD}$	oxygen demand from COD
$P_{A,in}$	inlet pressure, bar or kPa
$P_{A,out}$	outlet pressure, bar or kPa
Q	liquid flow rate, m^3/h
Q_A	aeration rate, Nm^3/h
Q_F	feed flow rate, m^3/h
Q_W	waste flow rate, m^3/h
R	recirculation ratio: recycle flow per feed flow
S_{COD}	COD substrate concentration
SAD_{bio}	specific aeration demand (relating to biological treatment), Nm^3/m^3
SAD_m	specific aeration demand (relating to membrane area), Nm^3/(m^2.h)
SAD_p	specific aeration demand (relating to permeate volume), Nm^3/m^3
$SEDA_m$	specific energy demand for membrane air scouring, see $E_{A,m}$
t	time, membrane life, plant life
t_c	chemical clean cycle time, hrs or d
t_p	physical clean cycle times, mins or hrs
V	tank volume, m^3
X	MLSS concentration, mg/L
y	depth of the aerator in the tank, m
Y	sludge yield, kgVSS per kg COD or BOD
Y_{obs}	observed sludge yield, kgVSS per kg COD or BOD
α	OTE correction factor for solids
β	OTE correction factor for salinity
γ	OTE correction factor for temperature
ΔS	change in concentration of substrate (COD, TKN or nitrate), kg/m^3
λ	ratio of substrate to MLSS concentration
λ_{COD}	biomass COD content, generally ~1.1 kg COD per kg MLSS
λ_{TKN}	biomass TKN content, TKN per g MLSS
k	dimensionless function of temperature
ε	blower efficiency, %
ρ_A	air density, kg/Nm^3
τ_c	chemical clean duration, mins or hrs
τ_p	physical clean duration, s or mins

1 Introduction

1.1 Industrial effluents

Industry accounts for about one quarter of all freshwater demand, and there are few industries which do not consume large quantities of water in generating specific products (Table 1-1). Given that only a limited amount of water is consumed by the industrial process, either through a change in phase (primarily to steam) or inclusion in the industrial product (such as beverages), it follows that a large amount of wastewater is generated by industrial activity. Such waters normally vary in quality temporally (seasonally and/or diurnally), as well as according to the application or duty, to a greater extent than that of municipal effluent (i.e. sewage). Moreover, in many cases industrial wastewaters tend to be more recalcitrant (i.e. biorefractory) than municipal ones, i.e. they are less readily treated biologically and the operating conditions have to be adjusted accordingly.

Table 1-1 Approximate water demand for various items (Waterfootprint, 2014)

Item		*Volume per unit*	*Water demand*
Material	Plastic	-	0.2 L/g
	Steel	100,000 L, car	0.31 L/g
	Bovine leather	300 L, leather jacket	17 L/g
Food	Beer	74 L, glass	0.3 L/g
	Paper	-	0.3-2.6 L/g
	Banana	160 L, large banana	0.8 L/g
	Wine	110 L, glass	0.88 L/g
	Milk	255 L, glass	1 L/g
	Eggs	200 L, egg	3.3 L/g
	Chicken	-	4.3 L/g
	Pork	-	6.0 L/g
	Cotton	2,700 L, T-shirt	11 L/g
	Beef	-	15 L/g
	Chocolate	1,700 L, bar	17 L/g
Power	Uranium	-	90 L/gJ
	Natural gas	-	110 L/gJ
	Coal	-	160 L/gJ
	Hydropower	-	22,000 L/gJ
	Biomass	-	70,000 L/gJ

As well as the discharged effluent, the quality of the influent water demanded by industry varies considerably from one duty to another. For some industrial processes the discharged water quality is not significantly lower than that of the feedwater. Cooling towers, for example, concentrate the water as a result of the evaporative cooling process, but do not add significantly to the pollutant content: the main impact is on the temperature. For most industrial sectors, however, there is a significant pollutant load resulting from their activity, demanding a level of treatment either for safe discharge to the environment or, increasingly, reuse within the process. It is opportunities for water recycling which have to some extent driven the uptake of "high-cost/high-value" process technologies, such as membrane separation, capable of providing reliably high treated water quality. On the other hand, the significant cost of these technologies combined with the challenging timeframe for return on investment usually demanded by most industrial sectors often mitigates against their implementation.

1.2 Industrial effluent treatment processes

The treatment of wastewater relies on a number of individual unit operations which are combined to make a process, or process treatment scheme. The unit operations themselves are fundamentally defined by the principles by which they work. Although process treatment technologies *per se* may be complicated and diverse in practice, their governing principles are largely limited to chemical, biochemical and physical processes (Fig. 1-1). Thus, for example, the simplest physical process is sedimentation, whereby particles are removed from water by settlement in large vessels. This process can be intensified by rotating the vessel, in effect enhancing the gravitational force, such as arises in hydrocyclones and centrifuges. The latter two technologies are very different in configuration to a sedimentation tank, but are nonetheless based on the same fundamental principle.

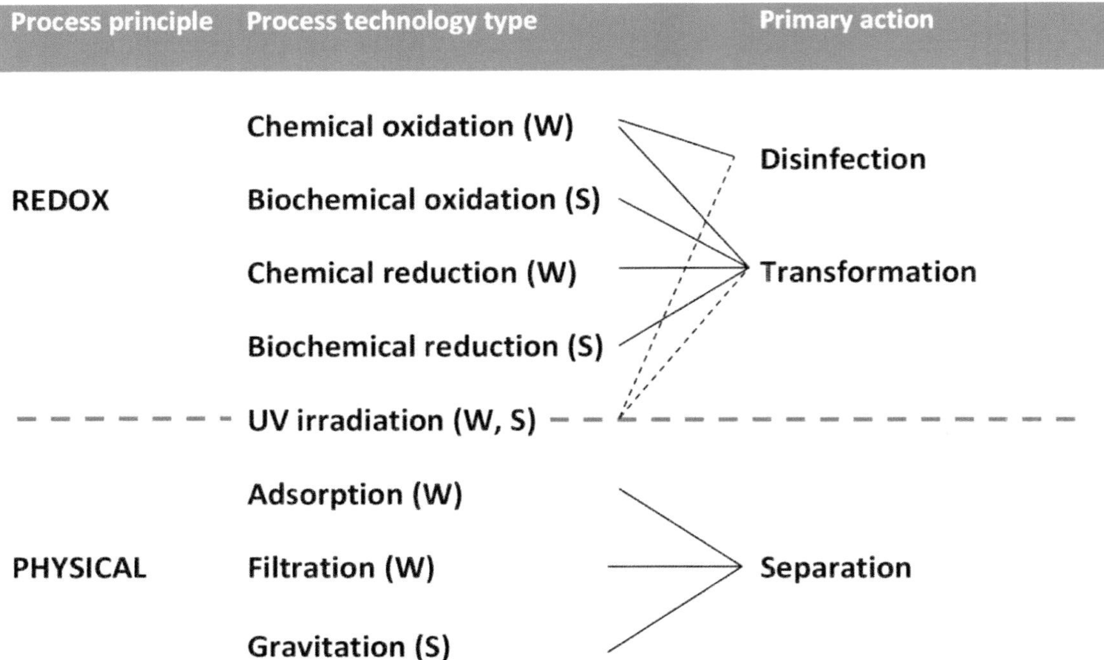

Figure 1-1 Process principles, process technology type, and primary action (W: water treatment; S: sewage treatment)

Chemical and biochemical oxidation processes achieve the same goal but employ different reagents and process configurations. Thus, the pollutant ammonia can be oxidised to nitrate either biochemically using micro-organisms (and specifically "nitrifiers") or chemically using chlorine – a very widely used chemical reagent in the water industry. The chlorination route is relatively rapid, in the region of 10-30 minutes, but leads to the formation of potentially harmful chlorinated by-products. The biochemical process is slow, taking anything between 6 and 48 hours depending on the nature of the feedwater, but is extremely efficient in terms of residual chemical by-products.

These types of reactions fall under the general term "redox", an abbreviation of "reduction/oxidation", since the oxidation of one species must necessarily be accompanied by the reduction of another for electroneutrality to be preserved. Thus processes based on chemical reduction (for example the quenching of excess chlorine using bisulphite) or biochemical reduction (the reduction of nitrate to nitrogen gas) are also examples of the redox principle. However, in the fields of microbiology and biochemistry such reactions are given specific names, for example the biochemical reduction of nitrate to nitrogen is termed "denitrification" and oxidation of ammonia to nitrate called "nitrification". There also exist many important non-redox chemical processes, such as pH adjustment or precipitation of alkaline earth salts (such as calcium carbonate or sulphate).

An exhaustive listing of all individual technologies based on the generic processes shown in Fig. 1-1 is beyond the scope of this book. There are dozens of different technology types, and these may act through more than one process principle. For example, chlorination can both oxidise and disinfect water. A simple slow sand filter combines surface filtration with biochemical oxidation, through the formation of a thick biofilm (or, as it is commonly referred to, "Schmutzdecke") on the filter surface. A membrane bioreactor combines the same two process principles, but the configuration of the technology differs completely from the slow sand filter.

Most biochemical (more usually termed biological) processes are intended to remove organic carbon, since this is the primary food source of micro-organisms generally. Such technologies (Fig. 1-2) can be roughly divided into two categories: "fixed film", where the biomass responsible for carrying out the biochemical reactions is affixed to some medium, and "suspended growth", where the biomass is distributed as particles (or "flocs") in a tank. Some technologies, such as the integrated fixed film activated sludge (IFAS), combine both of these aspects in their design. In this case plastic media are added to the biotank to encourage the growth of a fixed film on the media and thus improve the distribution and the retention of the biomass in the tank.

Biological processes can further be categorised according to the nature of the biology, and specifically the redox reaction (or electron transfer). The latter normally relates to the food source of the micro-organism concerned, which may then in turn depend on the prevailing conditions and, specifically, the presence of a source of oxygen. The conditions are thus generally defined according to whether there is a supply of dissolved oxygen or DO ("aerobic" conditions), an absence of DO but a supply of oxyanions like nitrate ("anoxic" conditions) or the absence of both of these ("anaerobic"). The conditions will then determine the biochemical conversion (Section 2.3.3).

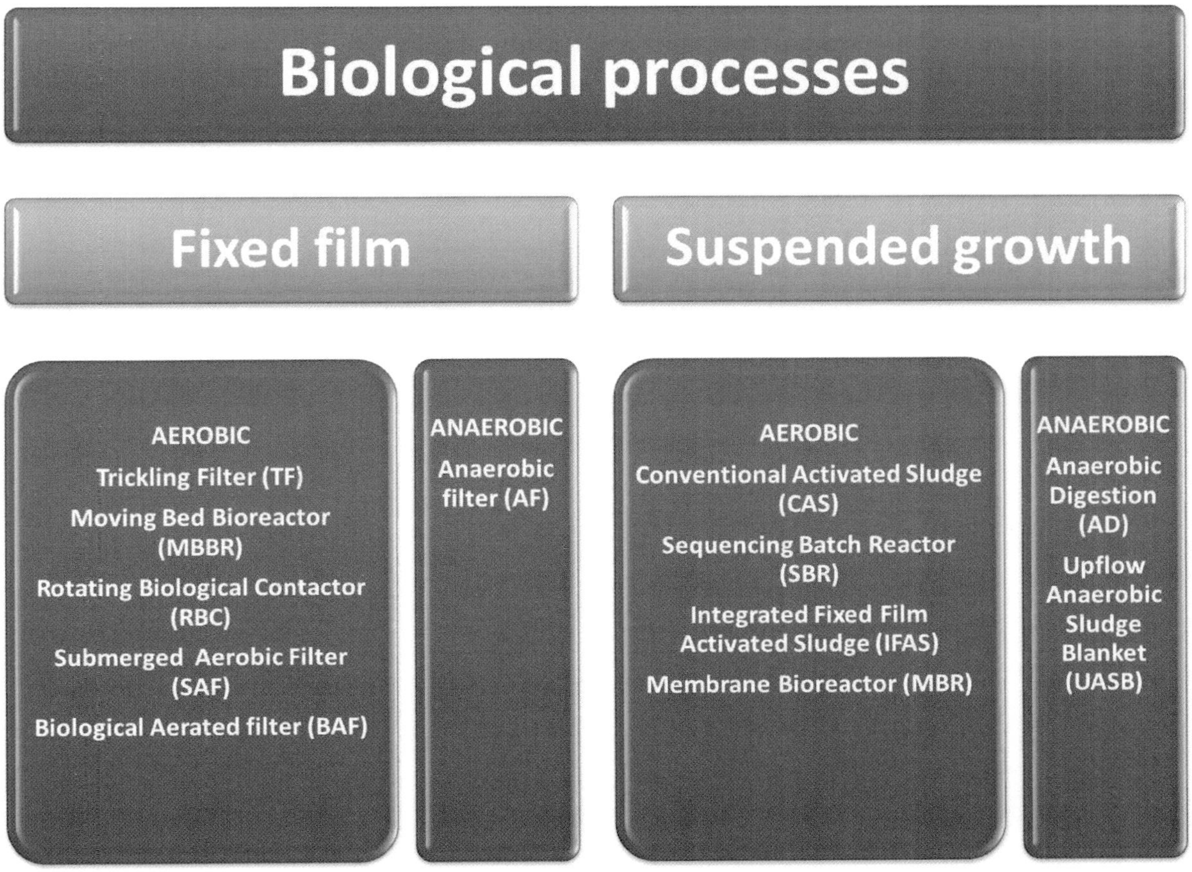

Figure 1-2 Types of biological process technology

1.3 *Membrane bioreactors*

1.3.1 The MBR technology

Membrane bioreactors (MBRs) are an example of a suspended growth process (Fig. 1-2); although other configurations have been studied they have yet to be commercialised. Whilst most MBR installations operate aerobically, there are examples of anaerobic applications; anaerobic treatment is generally more viable for wastes which have high levels of organic carbon, and is widely used for digesting wastewater sludge. Whilst various process adaptations of the MBR technology have been explored, such as the addition of media to the tank, the basis of the process – the use of a membrane to retain the biomass within a suspended growth biological process – has remained unchanged since its original development.

Membrane bioreactor technology is applied equally effectively to both municipal and industrial wastewater. Municipal MBRs have grown in size over the last decade to the extent that they are now being designed to treat over 350 MLD (megalitres per day), expressed as peak daily flow for municipal feeds, and the trend appears to be for increasingly large installations. Industrial MBRs, by the nature of their purpose, tend to remain relatively small – generally less than 10 MLD - and are designed for specialist applications (Chapters 3 and 5). Therefore, it may be expected for industrial effluent treatment plants not necessarily to increase in size but become more diverse in application, extending to increasingly challenging wastewaters. This tendency is reflected in the number of emerging specialist niche technology suppliers (Chapter 4).

1.3.2 MBR drivers

The drivers for the uptake of MBRs over other advanced wastewater treatment technologies can be summarised as being primarily (a) a requirement for high-quality treated water and (b) spatial restrictions. The process provides the highest quality treated water of all the biological treatment technologies in terms of residual concentrations of suspended inorganic and organic matter (particulates and colloids) and, in the case of municipal effluents, pathogenic bacteria and viruses.

MBRs are particularly attractive in instances where the treated water is to be desalinated by a pressure driven dense membrane process such as reverse osmosis (RO) or nanofiltration (NF). Since very significant removal of colloidal species is attained by the MBR, the RO/NF membrane fouling propensity of the product water – usually represented by the silt density index (SDI) – is very low. Other than the standard protection of the RO/NF membrane by a cartridge filter, no additional treatment of the MBR permeate is required for downstream membrane-based desalination. A number of such reuse plants exist worldwide where the recovered and desalinated water is used for industrial processes, such as steam-raising for electrical power generation, cooling, acid leaching operations or washing/laundering. The reuse of water for purposes where direct human contact is minimal reduces or eliminates the requirement for rigorous disinfection. Data from a recent review (GWI, 2012) suggests that the industrial effluent reuse/desalination market is growing by between 1.8% for the pulp and paper industrial sector to around 17% for the petrochemical and power generation sectors (Fig. 1-3).

The reduced size and footprint of MBR plants over that of conventional biological treatment technologies becomes important when (a) unit land costs are high and/or rapidly increasing, (b) space availability is limited, and (c) legal constraints have been imposed on the installation's visual impact. The latter has led to the housing of MBRs in bespoke buildings (Fig. 1-4) which can be demonstrably unlike those normally associated with conventional wastewater treatment. This particular facet of the technology has strongly influenced its implementation at a number of municipal sites globally, including subterranean installations such as those at Swanage in the UK and Guangzhou in China.

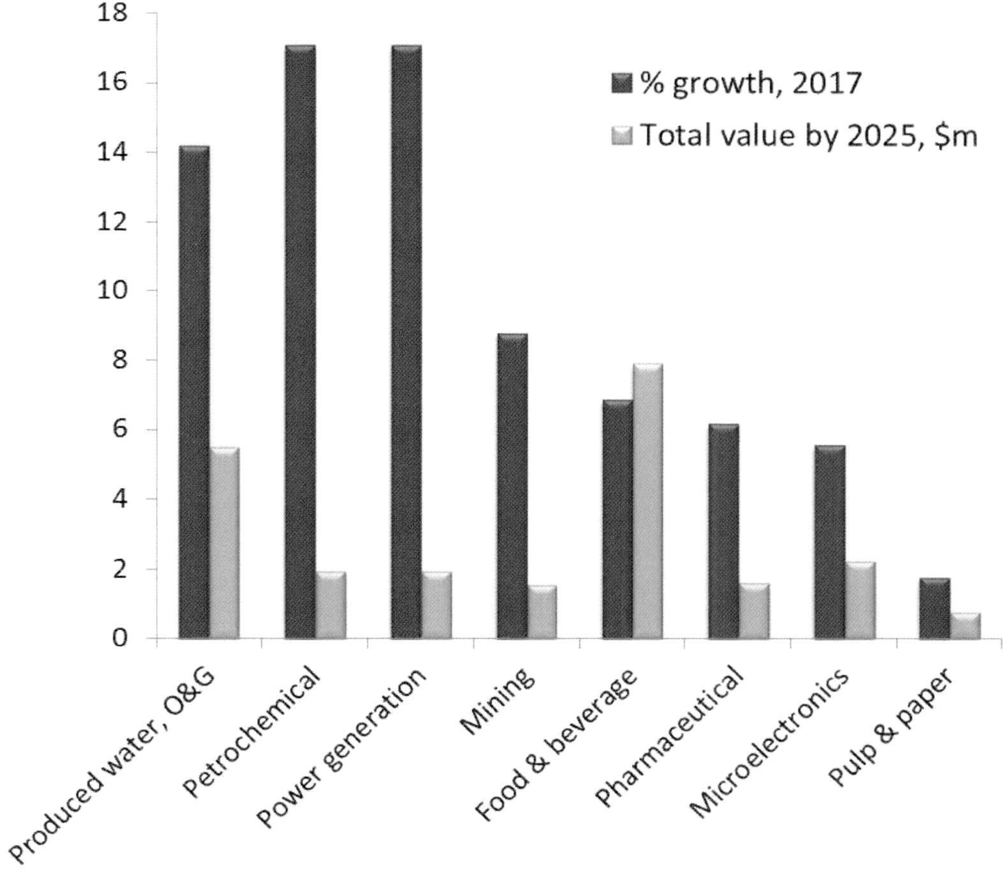

Figure 1-3 Market growth rate and value by industrial sector; data extracted from GWI (2012)

(a) (b)

Figure 1-4 MBR installations at (a) Porlock, UK, and (b) Busan, Korea

1.3.3 MBR barriers

The key barriers to MBR implementation are perceived as being cost and process complexity. Whilst the real cost of the membrane component has tended to decrease, particularly over the course of the 1990s when implementation dramatically increased, the energy demand is invariably higher than that of conventional processes due to the increased aeration demand (Section 2.7.2.1). Moreover, the operation of the membrane component is substantially more complicated in an MBR than in the secondary clarification process employed in the conventional process (Fig. 2-1(b)), requiring a chemical cleaning sequence to maintain the membrane permeability (Section 2.4.1). Whilst chemical cleaning is normally effective against membrane surface fouling, the same is not true of membrane channel clogging (Fig. 2-11). Clogging usually requires manual intervention to rectify the problem, further adding to the operational costs.

A key factor determining the overall process cost is the membrane life. This remains a challenging parameter to define, and it is generally assumed to be adversely affected by the application of the cleaning chemical. However, evidence from some of the most established plants suggests that the membrane life can exceed 10 years if the design parameters are appropriately selected and the membranes properly maintained.

Whilst operational costs remain higher than those for conventional treatment, whole-life costs (often calculated over a 25-year period) appear to be comparable if possibly sensitive to the assumed membrane life (Brepols et al., 2010). Improvements to energy efficiency continue to be made, and reducing energy demand remains a key focus of research and development. There is also increased interest in anaerobic MBRs which can potentially be considerably lower in carbon footprint for feedwaters relatively high in organic carbon – as is the case for many industrial effluents. Despite the challenges imposed by the technology, the MBR market (for both the industrial and municipal sectors combined) continues to grow at a compound annual growth rate (CAGR) of around 15%, bolstered by rapid expansion of water reuse (Section 2.1) in the Asia-Pacific region in particular. The ever-increasing size of the municipal installations (Table 2-1) appropriately reflects the increased confidence in the process along with the global requirement for higher-purity treated water, with the onus on conserving freshwater supplies in water-scarce regions. The continued wastewater market penetration of the technology would therefore appear assured.

In the following chapters, the elements of MBR design and operation are discussed, including costs (Chapter 2) and industrial elements by sector (Chapter 3). Commercially available MBR technologies and their characteristics are listed in Chapter 4, and industrial case studies (industrial MBRs in practice) are presented in Chapter 5.

References

Brepols, C., Schäfer, H. and Engelhardt, N. (2010). Considerations on the design and financial feasibility of full-scale membrane bioreactors for municipal applications. *Water Sci. Technol.*, **61**(10), 2461–2468.

GWI (2012). Industrial Desalination and Water Reuse: Ultrapure water, challenging waste streams and improved efficiency. ISBN: 978-1-907467-18-9.

Waterfootprint (2014) www.waterfootprint.org (accessed July 2014).

2 MBR design and operation fundamentals

2.1 Introduction

The term "membrane bioreactor" (MBR) is generally used to define those wastewater treatment processes where a perm-selective membrane (eg microfiltration/ultrafiltration, MF/UF) is integrated with a biological process (a suspended growth bioreactor, Fig. 2-1a). This is to be distinguished from a "polishing" process where the membrane is employed as a discrete tertiary treatment step with no return of the active biomass to the biological process (Fig. 2-1b). All commercial MBR processes available today are of the integrated type (Fig. 2-1a) and utilise the membrane as a filter, rejecting the solid materials which are developed by the biological process, thus resulting in a clarified (and substantially disinfected) effluent product.

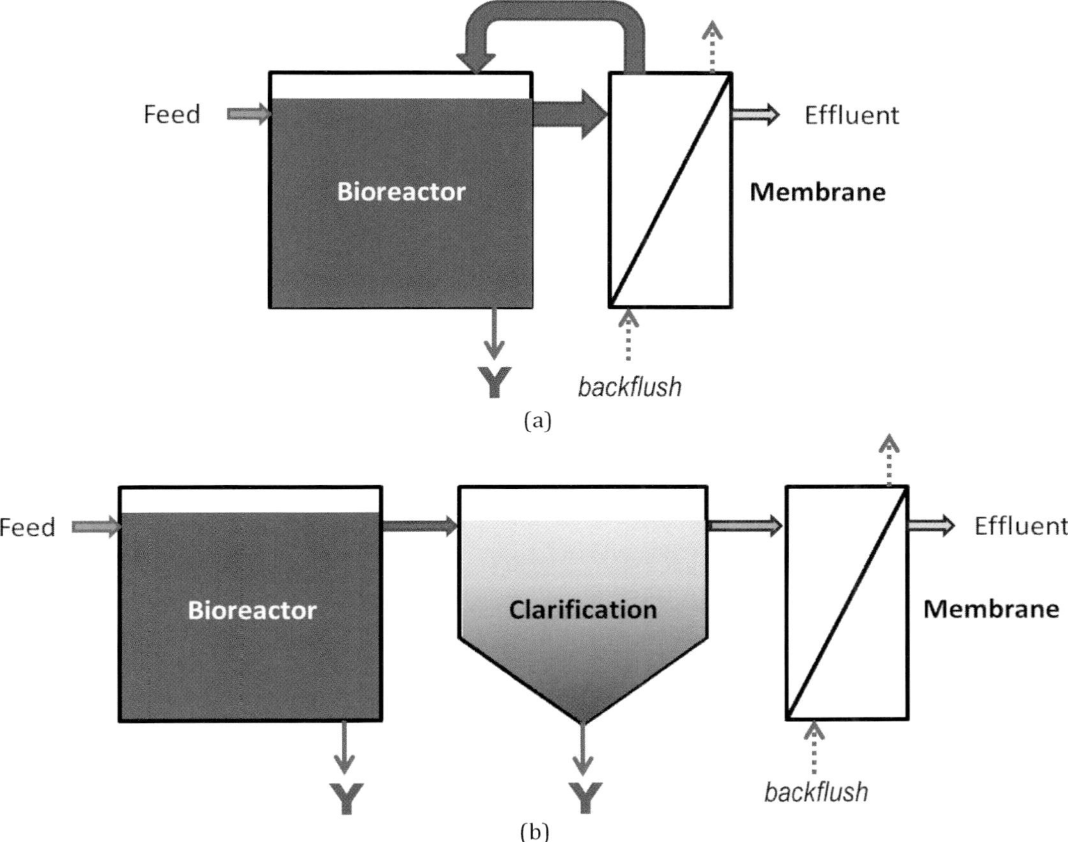

Figure 2-1 (a) Membrane bioreactor vs. (b) membrane polishing for wastewater treatment

An MBR is essentially a version of the conventional activated sludge (CAS) system. Whereas the CAS process employs a secondary clarifier (or sedimentation tank) for solid/liquid separation, an MBR uses a membrane for this function. This provides a number of advantages relating to process control and product water quality. Firstly, the sludge solids (or mixed liquor suspended solids, MLSS) are completely retained in the bioreactor, such that the solids retention time (SRT) in the bioreactor can be completely controlled regardless of the hydraulic retention time (HRT). This is to be distinguished from a CAS, where the flocculant solids (or flocs) that make up the biomass have to be allowed to grow in size to the point where they can be settled out using the secondary clarifier. This means that in a CAS the HRT and SRT are related, or coupled, since the floc size, and thus settlability, relates to the HRT. Secondly, the small pore size (<0.5 μm) of the membrane means that the treated effluent is of very high clarity and significantly reduced pathogen concentration. MBR processes can thus provide a substantially clarified and disinfected effluent of high enough quality to be discharged to sensitive receiving bodies or to be reclaimed for such applications as urban irrigation, utilities or toilet flushing. Thirdly, the retention of the solids and increase in SRT also impacts on both the plant footprint and the volumetric sludge waste generation, both of which are smaller than for a CAS process. These arise because the flocculant solids concentration in the bioreactor is higher, such that the same quantity of solids is retained in a smaller volume. Lastly, the longer solids retention times tend to provide better biotreatment overall. The conditions encourage the development of the slower-growing micro-organisms, and specifically the nitrifiers which convert ammonia into nitrate (Section 2.3.3). MBRs are thus especially effective at nitrification.

MBRs are thus becoming increasingly favoured for water and wastewater applications where a high treated water quality is required – especially for reuse – and where space is limited. All geographical regions appear to be experiencing a comparatively high rate of expansion (expressed as compound annual growth rate, CAGR) of the MBR market, with reference to the GDP of the country in which the technology is being installed, and significantly so in China. MBRs are now implemented in more than 200 countries and global market growth rates of up to 15% are regularly reported in various market analyses (BCC, 2008; MarketsandMarkets, 2013, Frost and Sullivan, 2013), although growth rates and the extent of implementation vary regionally according to the state of economic development and infrastructure. Confidence in the process appears to be increasing as the number and size of (predominantly municipal) reference installations grows, with a number of plants over 100 MLD in capacity (expressed as peak daily flow, or PDF) now installed (Section 2.6).

Whilst the market is dominated by a few established global players, the range of membrane products continues to expand with well over 60 MBR membrane module products available. The market value for the membrane equipment portion of MBR facilities was estimated at $0.75b in 2011, with around 40% of the revenue going to the Asia Pacific region. The total installed capacity of MBRs of both industrial and municipal installations in 2017 is expected to exceed 12,000 MLD (MarketsandMarkets, 2013).

2.2 Membrane terms and materials

A membrane as applied to an MBR is simply a material that allows water and dissolved matter to pass through it whilst rejecting solid particulate materials. It is thus perm-selective, with the degree of selectivity depending on the membrane pore size. In the case of MBRs, the range of pore sizes is relatively small, ranging from the microfiltration (MF) range of 0.1-0.4 μm to the coarse ultrafiltration (UF) range of, generally, between 0.02 and 0.1 μm. More than half the available commercial products have a rated pore size between 0.08 and 0.1 μm (Santos et al., 2011)

There are two different types of membrane material: polymeric and ceramic. The membrane material, to be made useful, must then be formed (or configured) in such a way as to allow water to pass through it; three membrane configurations apply to MBRs generally (Section 2.3.1).

A number of different polymeric and ceramic materials are used to form membranes. Membranes generally comprise a thin surface layer giving the required perm-selectivity on top of a more open, thicker porous support which provides mechanical stability. Membranes are usually fabricated both to have a high surface porosity, or % total surface pore cross-sectional area, and narrow pore size distribution to provide as high a throughput and selectivity as possible. The membrane must also be mechanically strong (i.e. to have structural integrity), and so firmly affixed to the supporting layer. Lastly, the material will normally have some resistance to thermal and chemical attack, that is, extremes of temperature, pH and/or oxidant concentrations that normally arise when the membrane is chemically cleaned, and should ideally offer some resistance to fouling.

Whilst ceramic membranes are more robust in terms of resistance to fouling and chemical attack, they remain limited to niche applications in MBR technology – primarily due to their relatively high cost. Ceramic multi-channel "monoliths" have found use in some applications, and a few ceramic flat sheet configurations have been introduced. However, commercial MBR membrane products are almost all polymeric.

More than half of the MBR membrane module products offered are based on polyvinylidene difluoride, or PVDF (Section 4.5), and the next most common material is polyethersulfone (PES). The combination of good chemical resistance and surface structure has meant that these polymeric materials dominate. However, polyolefinic membranes (in particular polyethylene, PE) are also used. The polyolefinic hollow fibre (HF) membranes are amongst the lowest in raw production cost of all MBR membrane materials, the pores being generated by extruding that material under controlled conditions (or "dry spinning") to produce slit-like pores. This slit-like pore structure is to be distinguished from the more complex thermal induced phase separation (TIPS) process for PVDF, which produces the more classic pseudo one-dimensional pores. The remaining materials - polyacrylonitrile (PAN), polysulphone (PS), polyvinyl alcohol (PVA) and polytetrafluoroethylene (PTFE) – are much less common.

2.3 Design

MBR design largely relates to the configuration of (a) the membrane, (b) the membrane separation process, and (c) the biotreatment process. Membrane configuration concerns the geometry of the membrane and the direction in which the water flows through it. The membrane separation process configuration concerns the placement of the membrane module in the overall MBR process. The biotreatment process configuration determines the biochemistry and, therefore, which pollutants are removed (organic carbonaceous materials, ammoniacal compounds and nutrients). Further elements of design comprise pre-treatment (screening and, for some applications, clarification) and post-treatment (generally either disinfection or desalination).

2.3.1 Membrane configurations

MBRs are generally grouped into three technology configurations - flat sheet (FS), hollow fibre (HF) and multi-tube (MT). Each of the configurations has tended to be better suited for particular applications, so the choice of configuration can be important when considering selecting MBR technology for a particular use.

2.3.1.1 Flat sheet membranes

Flat sheet (FS) membranes have planar configuration and are mainly rectangular (Fig. 2-2a), though other geometries exist for membrane modules designed to rotate. The element may be referred to as a "sheet", "cartridge" or, most commonly, a "panel".

Most of the rectangular panels are (semi-)rigid, though some of the thinner products are flexible and are mounted on a rigid frame. The rigid panels have a plastic backing plate to which the membrane's edge is welded on both sides. Water flows from outside to inside the panel and the permeate is collected either from the permeate outlet tube(s) or side/central manifold(s). Sludge is air-lifted up through the membrane channels, formed between adjacent panels in a cassette (Fig. 2-3a), by air bubbles provided from a coarse-bubble aerator placed underneath the module. The modules may be stacked to provide a double deck, with further stacking possible for a few products. Flat sheet membranes are used almost exclusively for immersed MBRs for both industrial and municipal applications, where they are sometimes favoured for smaller installations on the basis of their operational simplicity.

2.3.1.2 Hollow fibre membranes

Hollow fibre (HF) membranes are normally vertically oriented with the aerators either integrated with the module or fitted to the frame. HF elements (Fig. 2-2b) are most commonly referred to as a "module". The fibres are usually provided with some slack, to allow them to move laterally in the flow of air bubbles, which also air-lift the sludge solids through the fibres.

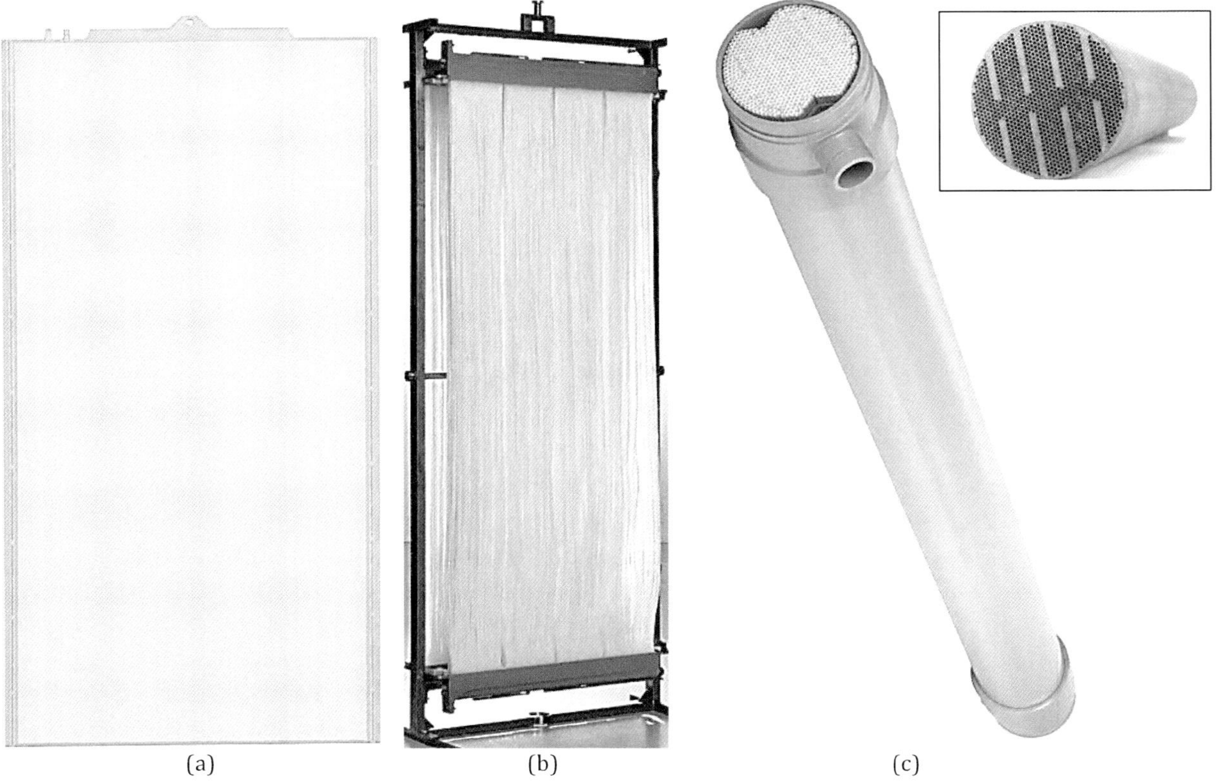

(a) (b) (c)

Figure 2-2 MBR membrane configurations, single elements: (a) flat sheet (FS), (b) hollow fibre (HF) and (c) multi-tube (MT). Inset is an example of a ceramic multi-channel (MC) monolith

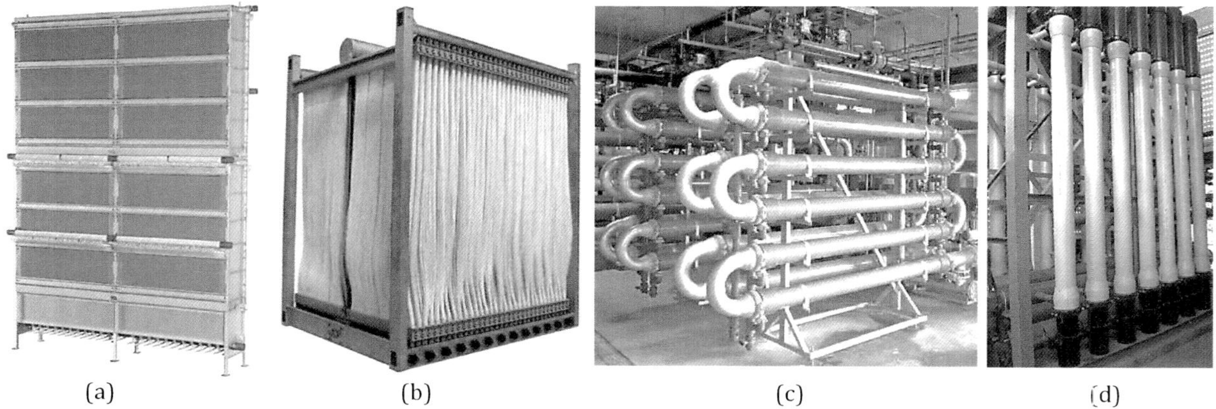

(a) (b) (c) (d)

Figure 2-3 MBR membrane configurations, stacks/racks: (a) FS cassette (or unit or stack) (b) HF cassette (or stack), (c) pumped MT skid (or rack), and (d) air-lift MT skid

Water flows from outside to inside the individual fibres and is collected from the fibre potted either at one or both ends and fed into a collection chamber or manifold. For some products the fibres are reinforced with a braided core to which the membrane is bonded. HF membranes are used almost exclusively for immersed MBRs, both for industrial and municipal applications, where they are often favoured for larger installations on the basis of their lower membrane aeration demand (Section 2.3.5.2).

2.3.1.3 Multi-tube membranes

Multi-tube (MT) membranes are currently the only MBR membrane products which are standardised, and – for MBR technologies – are only used for sidestream configurations. A module (Fig. 2-2c) comprises a bundle of tessellated tubes within a standard-sized cylindrical casing – most often 8" (200 mm) in diameter – and the water flows from inside to outside the membrane tubes. There also exist ceramic membranes which comprise a single block of material, or a "monolith", containing channels (hence multi-channel or MC) of either cylindrical or square geometry (Fig. 2-2, inset).

For pumped systems, the modules are placed in series to provide a high conversion (Section 2.4.1), with one individual stream comprising a number of horizontal "loops" (normally 2-4 in all) in a serpentine arrangement within a skid (Fig. 2-3c). Pumped systems are often favoured for treating small effluent flows from industrial installations on the basis of their robustness and operational flexibility and control. The air-lift type competes with FS and HF immersed systems. For air-lift sidestream systems, the modules are discrete, vertically-oriented and generally longer than those used for the pumped type (Fig. 2-3d).

2.3.2 Membrane separation process configurations

Two process configurations exist for the membrane: immersed (iMBRs) and sidestream (sMBRs). The immersed configuration is also referred to as "submerged", which obviously leads to confusion when abbreviated, or "internal" (as opposed to "external" for sidestream membranes). Further process configurations, based on the way in which the membrane is used, include "diffusive" and "extractive". However, neither of these have reached the stage of commercialisation: all current commercial MBRs employ a membrane solely for retaining the MLSS in the bioreactor whilst producing a high-quality permeate product.

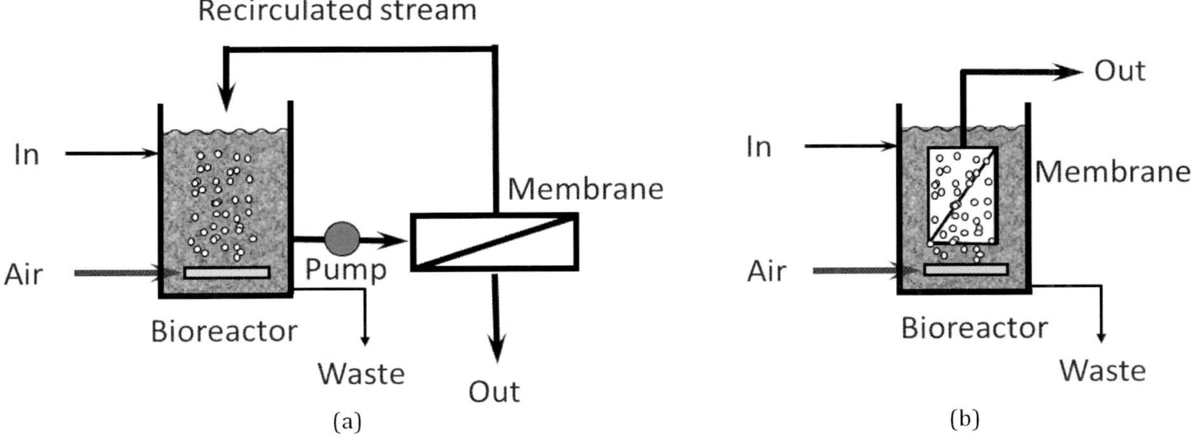

Figure 2-4 Membrane bioreactor process configurations: (a) sidestream, and (b) immersed

The first MBRs, developed in the late 1960s, were based on the sidestream configuration (Fig. 2-4a). For this configuration the membrane system is placed outside the biological treatment process tank. The sludge is then pumped through the membranes in "crossflow" mode, normally at an elevated pressure (2-4 bar), and the permeate forced through the membrane under the combined action of the pressure and the crossflow (at a liquid velocity of 2-4 m/s). The unpermeated stream is then returned to the biotank. For the more recent immersed configuration (Fig. 2-4b), which was introduced in the early 1990s, the membrane is placed in the biotank and scoured with air introduced by an aerator placed beneath or within the membrane module. As with almost all membrane separation processes, key to maintaining the flow of water through the membrane is the generation of "shear". This relates to the flow of fluid over the membrane surface, which for sidestream configurations is sustained by the flow of sludge and for immersed systems by the action of the air bubbles.

2.3.3 Biotreatment process configurations

The dissolved organic contaminant removal achieved by any biological process is dependent on whether oxygen is present, and in what form. Aerobic processes require dissolved oxygen (DO) to convert organic carbon to carbon dioxide and ammonia to nitrate (nitrification).

In the absence of DO, organic carbon may still be oxidised if an alternative source of oxygen is present – from an oxyanion and specifically from nitrate (NO_3^-). This then presents a means of removing nitrate, a nutrient, from the wastewater. Nitrate formed from nitrification of ammonia can be converted to nitrogen gas by biochemical reaction with organic carbon, a process known as "denitrification". This can only take place if dissolved oxygen is absent but nitrate is present – conditions described as "anoxic". Returning the treated nitrate-rich sludge to an anoxic zone in front of the aerobic treatment step thus allows the nitrate to be removed, a process modification known as the "Modified Ludzack-Ettinger" (MLE) process (Fig. 2-5). A combination of aerobic nitrification and anoxic denitrification permits most of the "total nitrogen" (TN) to be removed. TN removal is thus achieved at the expense of additional energy, required for transferring sludge between the two regions, but with slightly reduced oxygen demand (and thus biological process aeration energy) since part of the oxygen requirement is effectively provided by the nitrate (Section 2.7.2.3).

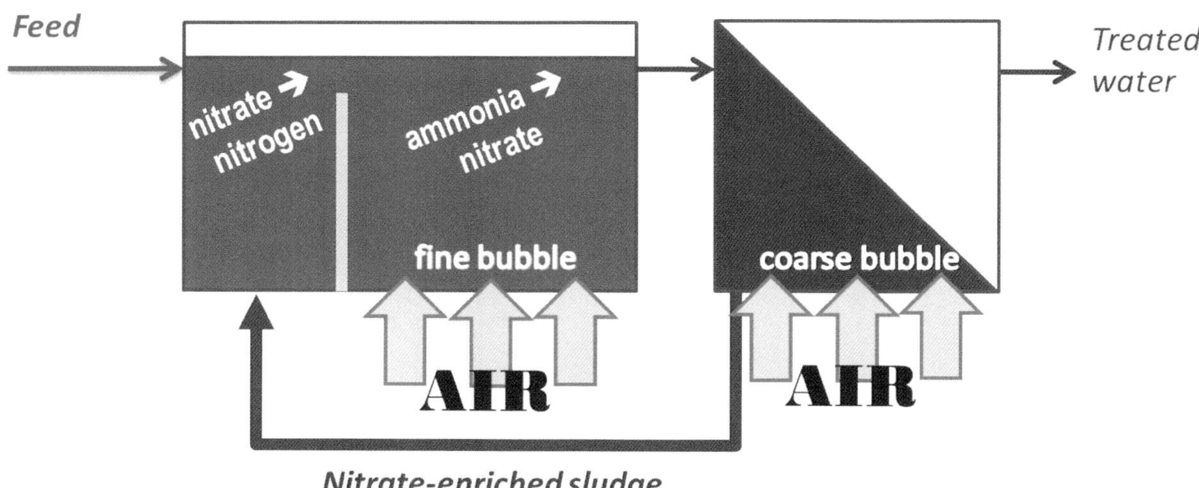

Figure 2-5 The Modified Ludzack-Ettinger MBR biological process configuration for denitrification

If there is no source of oxygen to allow any biochemical reaction with the organic carbon to take place – either from oxygen or oxyanions – then the conditions are described as "anaerobic". In the treatment of municipal wastewater, anaerobic conditions are applied as a treatment step in assisting the removal of phosphate from the water. Biological phosphorus removal (BPR) in this manner proceeds through recirculation of the sludge to an anaerobic zone (Fig. 2-6). This promotes phosphate removal in two ways: (i) by encouraging the phosphate to be taken up by the biomass (the microbiologically active component of the mixed liquor), and (ii) releasing the biological phosphate as inorganic phosphate, which can then be removed by chemical precipitation. Biological phosphorus (P) removal thus exerts further energy demand for sludge transfer to the anaerobic tanks, as well as chemical demand for the precipitation.

The above process configurations refer to the activated sludge process and have been widely applied to MBRs at full-scale for nutrient removal. Studies of the incorporation of other biological process technologies (Fig. 1-2) into MBRs have been reported. Hybrid MBR processes investigated have thus far included sequencing batch reactors (SBRs) (Scheumann and Kraume, 2009), moving bed bioreactors (MBBRs) (Ødegaard et al., 2012) and the integrated fixed film activated sludge process (IFAS) (De la Torre et al., 2013). However, such hybrid processes appear not to have reached the stage of full commercialisation.

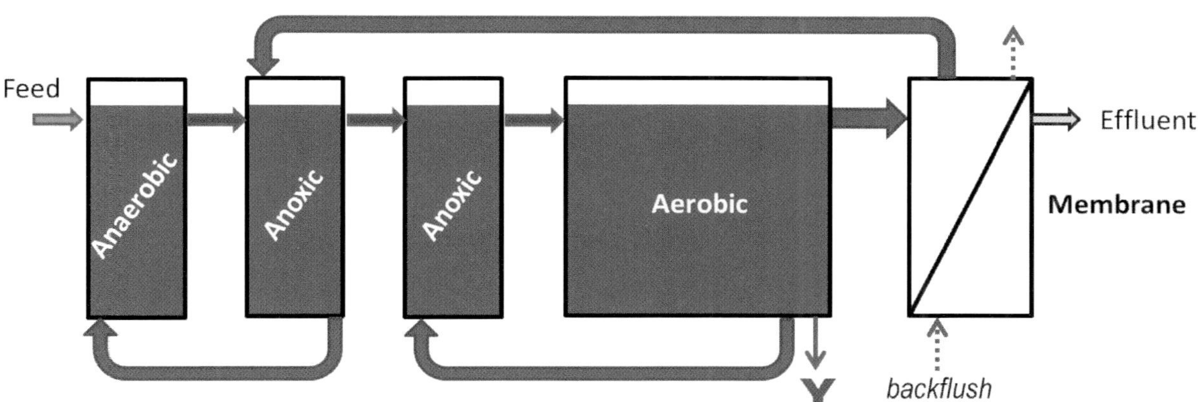

Figure 2-6 Biological phosphate removal, possible biological configuration

2.3.4 Anaerobic MBRs

Whilst almost all MBR technologies implemented are aerobic, there has been a surge of interest in anaerobic MBRs since the mid 2000s. The technology provides the potential for removing COD with a net energy benefit from the methane generated, albeit without nutrient removal. Interest within academia is evidenced by the publication of five independent reviews of the subject in various learned journals in 2012 alone (Dereli et al., 2012; Singhania et al., 2012; Skouteris et al., 2012; Smith et al., 2012; Visvanathan and Abeynayaka, 2012). Whilst the sidestream configuration (AnsMBR) was originally commercialised in the early 1990s and is still provided by at least one multinational company (Section 4.3.4), the most recent interest has been associated with the immersed configuration.

Available information suggests that AnsMBRs can provide a COD rejection of 99% or more, the % removal increasing with increasing feedwater concentration, and achieve a flux of 15-30 LMH for a range of food effluent applications. Operating conditions (CFV and TMP), when reported, appear to be similar to those employed for an aerobic sMBR, with a reduction in the CFV producing a corresponding reduction in the sustainable flux. This being the case, the value offered by anaerobic as opposed to aerobic treatment by an sMBR is determined by the balance of (from the perspective of the anaerobic option):

1) The OPEX benefit of the methane generated, which is then proportional to the difference in the feed and permeate COD concentration;
2) The OPEX benefit of the reduced process aeration (assuming all other aspects of the anaerobic and aerobic biological process OPEX to be similar);
3) The OPEX benefit of the reduced sludge production;
4) The OPEX penalty of the increased specific energy demand for the membrane filtration (which is proportional to the flux);
5) The CAPEX penalty associated with the larger membrane area demanded by the lower flux;
6) The overall cost penalty of supplementary downstream nutrient and residual COD removal, if required.

Since flux does not appear to be a function of loading, the anaerobic MBR option – as with the classical treatment – becomes more viable at higher loadings. This arises from a combination of the calorific value (CV) of the methane generated (1) and the reduction in process aeration (2), both of which are roughly linearly related to the COD (as is the proportional reduction in sludge (3)). The OPEX penalty (4), for a pumped sMBR, roughly equates to the permeability (Section 2.4.1), whereas the CAPEX penalty is inversely proportional to the flux.

It is because of the significant OPEX penalty that there has been recent interest in the immersed configuration (the aniMBR), which demands a much reduced energy for permeation and for which scouring can potentially be provided by the generated biogas. Pilot-scale studies of this configuration, along with data from a full-scale installation (Section 5.1.18), suggest that aniMBR fluxes are generally in the range of 4-10 LMH (Christian et al., 2010; Dereli et al., 2012; Diez et al., 2012; Robles et al., 2012) depending on the feedwater quality. There is also some indication from iHF studies that backflushing may significantly increase the sustainable flux (Diez et al., 2012; Robles et al., 2012).

An alternative to the classical suspended growth aniMBR (i.e. the anaerobic analogue of the aerobic process) is the combination of membrane separation with a UASB. In this case the membrane is challenged only with the supernatant of the UASB, the bulk of solids being retained in the bioreactor in granular form, and as such it is not a classical MBR process with an integrated biomass-separating membrane. This configuration, in which the membrane acts more as a polishing unit than as an integral part of the bioreactor, has undergone demonstration-scale trials at the Singaporean Public Utility Board's (PUB) Olu Pandan plant.

2.3.5 Aeration

2.3.5.1 Biological treatment

The oxygen demanded for biological treatment, often referred to as the "process" aeration, is dependent primarily on:

a) the inlet and outlet biodegradable organic carbon and Kjeldahl nitrogen levels, total Kjeldahl nitrogen (TKN) being total biologically oxidisable nitrogen, and
b) the rate at which biomass is created.

The absolute difference between mass flow rates of the inlet and outlet substrate levels, the substrate being the biomass "food" of biodegradable carbon and nitrogen, defines the load for each substrate. Since in wastewater applications the organic carbon is conventionally measured as "oxygen demand", rather than directly as total organic carbon (TOC), the oxygen demanded for organic carbon removal is normally taken as being represented by the chemical oxygen demand (COD) of the feedwater. For nitrogen oxidation (nitrification) the ratio of oxygen required to TKN is 4.57:1 on a mass basis (or 2:1 on a molar basis of O_2 to N). The oxygen mineralises the organic matter: organic carbon is converted to bicarbonate (HCO_3^-) and organic nitrogen to nitrate.

Since some of the organic carbon and nitrogen form new cells, generating more biomass, this proportion of the COD and TKN does not demand oxygen. The oxygen demand for the biological process is then determined by the difference between the load and the concentration of cells formed. The generation of biomass demands that these solids be continuously removed from the bioreactor to maintain a constant concentration of the biomass, referred to as the mixed liquor suspended solids (MLSS) concentration, in the bioreactor. This then forms a waste stream from the process, known as the "waste activated sludge" or WAS (Section 2.4.2).

In many cases further biological treatment is required for nutrient removal (nitrate and phosphate, normally denoted N and P) by using the MLE and BPR biological process configurations (Section 2.3.3). These modifications further impact on the oxygen demand, but do not significantly impact on the calculated oxygen requirement based on a simple mass balance approach (Section 2.7.2.1).

The actual aeration rate required to deliver sufficient DO to sustain biological treatment is dependent on key design parameters, such as the aerator type and the depth to which it is submerged in the tank, and the residual DO and MLSS concentration. The efficiency of transfer of oxygen from air into the mixed liquor is defined by the alpha factor (α), which is simply the ratio of the mass transfer coefficient of water to that of the mixed liquor. α decreases with MLSS concentration according to trends which appear to vary widely between different studies but which generally follow either a linear or pseudo-exponential relationship (Fig. 2-7). Thus, according to some relationships, the aeration rate required to maintain a specified DO residual may double between an MLSS concentration of 7 and 12 g/L.

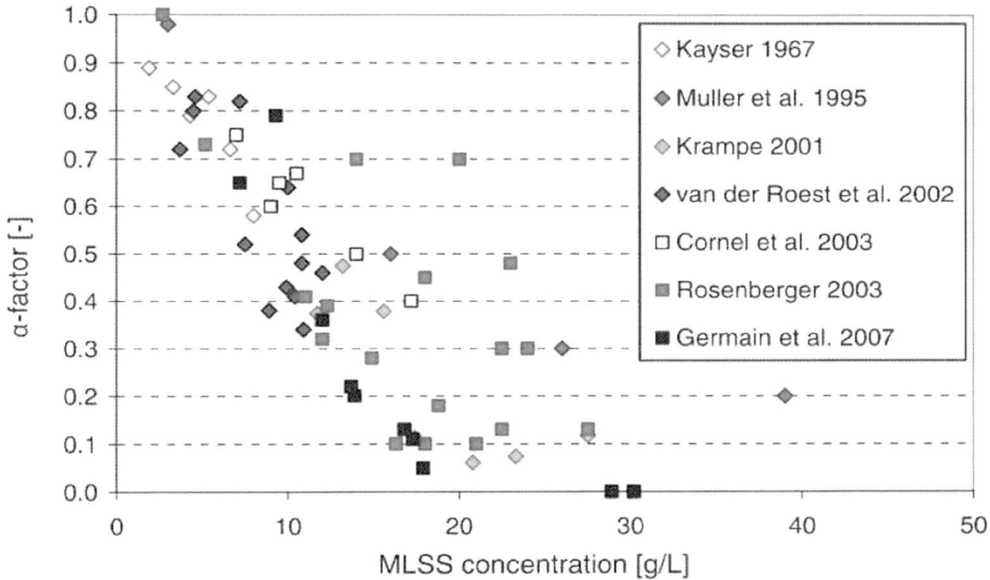

Figure 2-7 α-factor vs. MLSS (Henkel et al., 2011)

The exact nature of the relationship between α and MLSS concentration is highly dependent on the aerator type, the precise characteristics of the sludge and the aeration rate. Whilst the transfer of oxygen into the biomass increases with total bubble surface area, and thus with decreasing bubble size for the same flow, the bubbles tend to coalesce to form larger bubbles. The latter effect increases with decreasing bubble size and increasing air flow rate (and thus larger bubble population density). Moreover, the permeability of the diffuser increases with decreasing pore size – analogous to a membrane – increasing the pressure required to achieve a target air flow rate. Lastly, whilst mass transfer is promoted by small bubbles, the bubbles must also be capable of providing a degree of mixing of the sludge in the aeration tank which is best provided by larger bubbles. There is thus a practical lower limit of fine bubble diffuser (FBD) pore size. Coarse bubble aerators, on the other hand, are less prone to bubble coalescence due to a combination of the reduced bubble density and greater turbulence associated with them.

2.3.5.2 Membrane air scouring

Membrane aeration in immersed configurations is normally via coarse bubble aerators positioned beneath the membrane units (Fig. 2-8a,b,d,e), or sometimes integrated with the module (Fig. 2-8c). Aeration applied to membrane scouring can be normalised to give the "specific aeration demand" (*SAD*). Normalisation can be either against the membrane area (SAD_m, taking units of $Nm^3/(m^2.h)$) or the permeate volume (SAD_p, Nm^3/m^3), a "Nm^3" being the volume of 1 m^3 of air at a standard temperature of 20°C. SAD_p is then directly proportional to the specific energy demanded for membrane air scouring $SEDA_m$ or $E_{A,m}$ (Section 2.7.2.1), since all other components of air pumping (namely the blower efficiency, aerator depth, aerator type, inlet air pressure, and ratio of air and water enthalpies) are constant for a specific installation. SAD_p is the ratio of SAD_m to the flux in consistent units, i.e. in m/h; energy efficiency is improved both by increasing the sustainable flux and decreasing the membrane air scour rate required to sustain this flux.

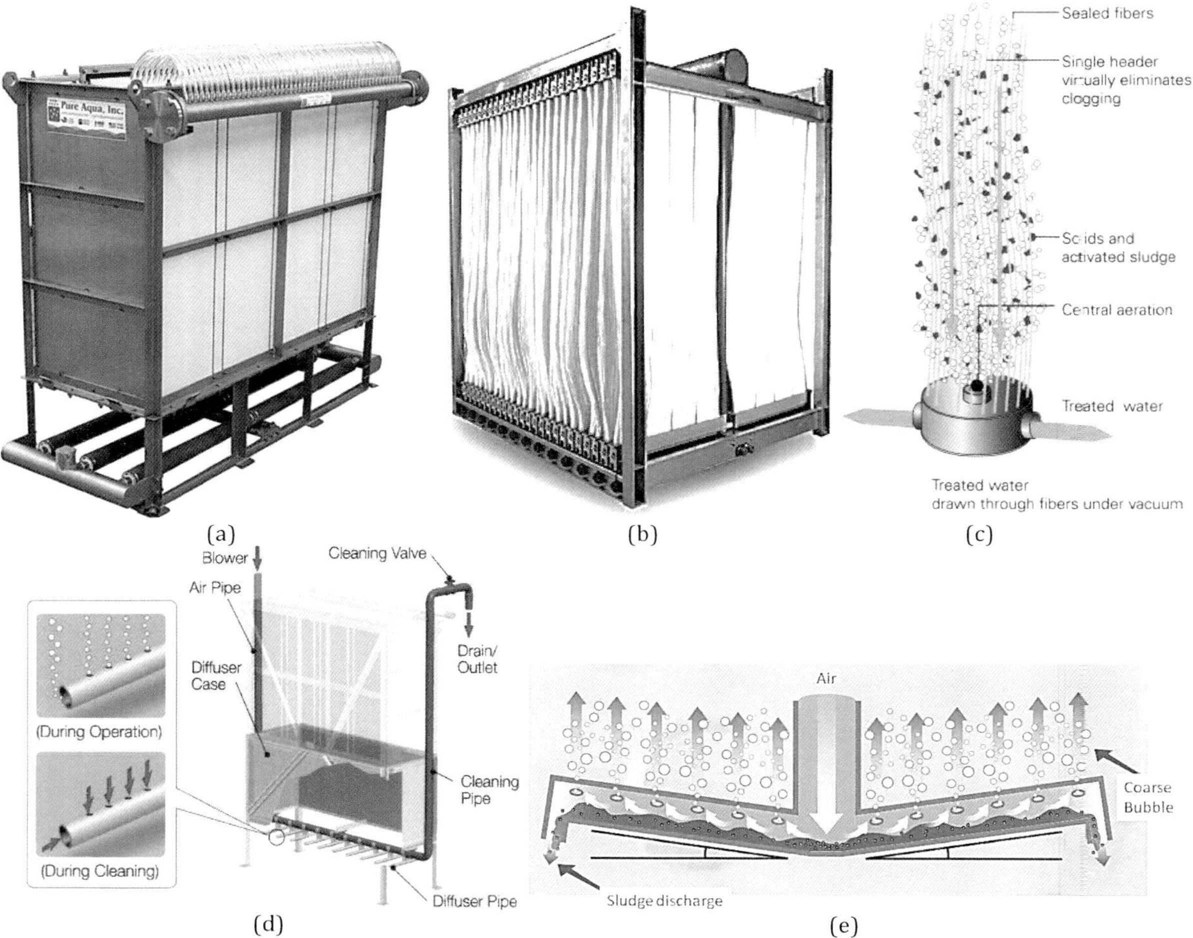

Figure 2-8 Flat sheet and hollow fibre aerators and membrane units: (a) Toray (with expandable rubber sleeve) (b) GE, (c) Koch Membrane Systems *PURON*® integrated aerator, (d) Kubota, with centipedal tube aerator, and (e) schematic of Econity membrane aerator

Whereas in the process tanks the aeration rate is determined by oxygen demand of the biodegradable organic matter, with the process designed to maintain a constant residual DO concentration, the aeration rate in the membrane tank is determined only by the membrane scouring requirement. The aeration demanded varies according to the membrane technology, both with regards to the total air flow and the mode of application. The latter can be either continuous or intermittent, with more intermittent aeration possible at lower fluxes (thereby maintaining the target SAD_p). In the case of the GE *Zeeweed* module, intermittent aeration can be for 10s in every 20s period ("10:10 aeration") or every 40s period ("10:30 aeration"), achieved by diverting the air flow between adjacent membrane trains.

Values for SAD_m range from 0.2 to 0.5 $Nm^3/(m^2.h)$, with iHF technologies tending to have slightly lower values than the iFS ones. The air-lift sidestream modules (Fig. 2-3d) have a similar range of SAD_m values to the iMBR technologies. Therefore, for target operational flux values typically around 20 LMH (or 0.02 m/h), the SAD_p is ~10-25 Nm^3 air per m^3 permeate.

2.4 Operation and maintenance (O&M)

The key process operational parameters can be summarised as follows:

Membrane side: Flux, pressure, permeability, recovery (conversion), rejection, membrane aeration rate or crossflow velocity, physical and chemical cleaning cycle times and protocols.

Biological side: Hydraulic and solids retention time, sludge recycle rate(s).

2.4.1 Membrane O&M

The flux (normally denoted *J*) is the volume of permeate generated per unit area of membrane per unit time (Fig. 2-9). This means that it takes SI units of $m^3/(m^2.s)$, or simply m/s, but it is normal to use non-SI units either m per day, litres per m^2 per hour (or LMH) or, in US units, gallons per square foot per day (GFD). MBRs generally operate at fluxes between 10 and 150 LMH, depending largely on the membrane and process configurations, and the feedwater quality.

The flux relates directly to the applied transmembrane pressure, or TMP in bar or psi (pounds force per square inch), which includes the hydraulic resistance offered by both the membrane and the interfacial region adjacent to it (or "fouling layer"). The ratio of the flux to TMP is then referred to as the "permeability" which normally takes the most convenient units of LMH/bar. In the US permeability is often termed "specific flux", and takes US units of GFD/psi.

Conventional pressure-driven membrane processes with liquid permeation can operate in one of two modes. If there is no retentate stream then operation is termed "dead-end" or "full-flow"; if retentate continuously flows from the module outlet then the operation is termed "crossflow" and applies to all sidestream systems. Crossflow implies that, for a single passage of feedwater across the membrane, only a fraction is converted to permeate product. This parameter – the ratio of permeate to feed flow – is termed the "conversion" or "recovery". The recovery is reduced further if product permeate is used for maintaining process operation, usually for membrane cleaning.

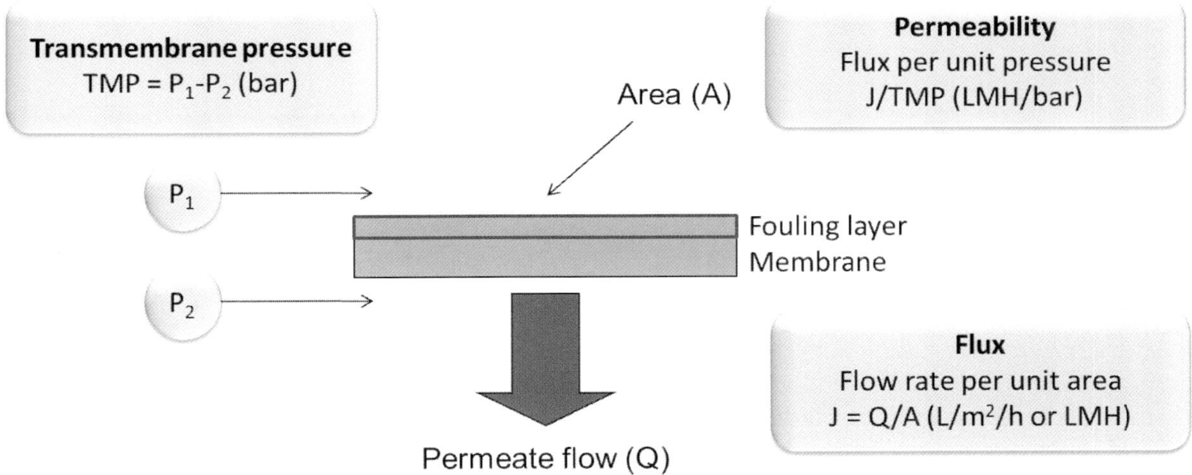

Figure 2-9 Flux, transmembrane pressure and permeability

Filtration always leads to an increase in the resistance to flow, normally termed "fouling". In the case of a dead-end filtration process, the resistance – manifested as an increase in the TMP – increases according to the thickness of the cake formed on the membrane. The increase in TMP is roughly proportional to the total volume of filtrate passed; for normal constant flux operation this means that the TMP increases rapidly with time, whilst permeability commensurately decreases. For crossflow processes, this deposition continues until the adhesive forces binding the cake to the membrane are balanced by the scouring force (the shear) of the fluid passing over the membrane. This then leads to much slower permeability decline than for dead-end operation (Fig. 2-10a).

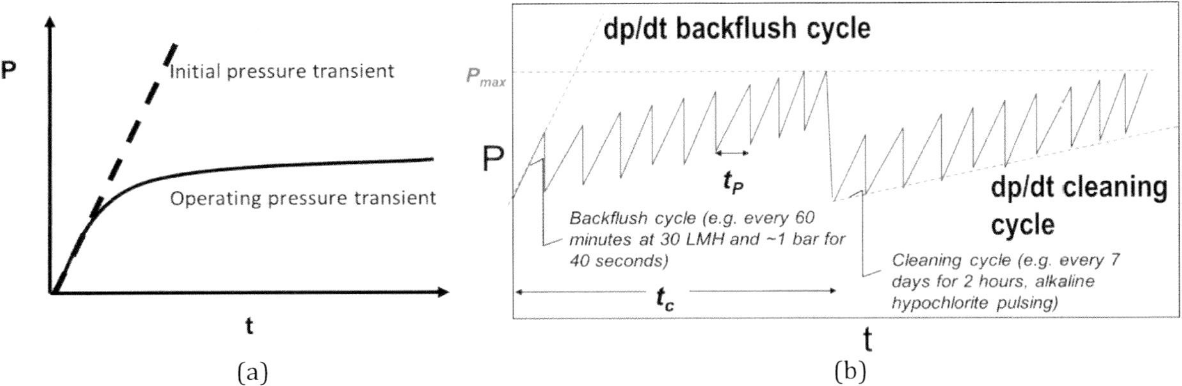

(a) (b)

Figure 2-10 Pressure transient during (a) crossflow, and (b) dead-end filtration

For dead-end processes, on the other hand, the membrane has to be cleaned to recover the permeability, producing a classic "sawtooth" pressure profile (Fig. 2-10b). Membrane cleaning can be by physical or chemical means, and both are employed in MBR O&M. Physical cleaning is normally achieved either by "backflushing", i.e. reversing the flow back through the membrane, or "relaxation" – ceasing permeation whilst continuing to scour the membrane with air bubbles (for an immersed process) or the crossflow (for a sidestream one). These two techniques may be used in combination, and backflushing may be enhanced by combination with air.

Physical cleaning removes gross solids attached to the membrane surface, generally termed "reversible" or "temporary" fouling, whereas chemical cleaning removes more tenacious material often termed "irreversible" or "permanent" fouling. Physical cleaning is less onerous than chemical cleaning in that it is generally more rapid, demands no chemicals, generates no chemical waste and is less likely to degrade the membrane. However, its effectiveness is limited, such that chemical cleaning is always required at some point. Since the original virgin membrane permeability is never recovered once a membrane is fouled through normal operation, there remains a residual resistance which can be defined as "irrecoverable fouling" which may build up over a number of years and so ultimately determine membrane life.

Chemical cleaning usually employs sodium hypochlorite, an oxidative chemical, in combination with mineral or organic acids (most often citric acid). These two chemicals are applied sequentially (i.e. one immediately following the other) with the acid clean often scheduled less frequently than the hypochlorite clean. Cleaning is normally conducted without removing the membrane from the tank or skid (and is hence termed a "clean in place" or CIP). If chemical cleaning is combined with backflushing this is referred to as a "chemically-enhanced backflush" (CEB). CEBs are routinely carried out on a weekly/monthly basis for iHF MBRs, and such protocols are normally referred to as "maintenance" cleans.

Chemical cleaning employed specifically to recover permeability demands higher reagent concentrations and longer contact times than maintenance cleans. Such "recovery cleaning" is not always effective – normally because the membrane has become clogged with sludge solids (Fig. 2-11a,b), a phenomenon often referred to as "sludging". In municipal wastewater treatment, membranes may sometimes also become clogged with "rags" (or "braids") formed from aggregated filamentous matter (specifically textile fibres such as cotton wool) in the feedwater. This is usually referred to as "ragging" or "braiding" (Fig. 2-11c). Sludging and ragging are the main causes of unscheduled manual intervention in municipal MBRs.

Membrane cleaning is usually the only operational parameter which is within the control of the operator. There is normally a maximum TMP (or minimum permeability, L) beyond which operation is undesirable due to the increased risk of irrecoverable fouling or clogging. The cleaning cycle fouling rate dP/dt (Fig. 2-10b) is thus critical in determining whether the maintenance cleans are sufficient for sustaining the target permeability and/or when recovery cleaning is required. Both physical and chemical cleaning then impact on the production rate, such that the net flux J_{net} is always less than the instantaneous flux J:

$$J_{net} = \frac{J}{t_p}\left(1 - \frac{\tau_c}{t_c}\right)\left(t_p - \tau_p - \frac{J_b \tau_p}{J}\right) \tag{1}$$

where J_b is backflush flux (normally 1-2 times the operating flux), t_p and t_c are the cycle times for the physical and chemical cleans, and τ_p and τ_c are the corresponding cleaning durations. Typically, for an HF iMBR, backflushing of 30-60 s duration (τ_p) is applied every 10 minutes (t_p) and maintenance cleaning scheduled weekly (t_c) and taking roughly two hours (τ_c).

(a) (b) (c)

Figure 2-11 Clogging of (a) flat sheet panel and (b) hollow fibre module. (c) Ragging at base of cassette

2.4.2 Biological process O&M

Compared with the membrane side, the O&M of the biological process is relatively simple, since there are actually few controllable variables. The two most important operating parameters are the hydraulic and solids retention times (HRT and SRT), of which only one (the SRT) is within the control of the operator given that the influent flow cannot normally be controlled. The SRT then determines the MLSS, which increases with increasing SRT, which then further impacts on both the oxygen transfer (Section 2.3.5) and the membrane separation process (Section 2.4.1). It also determines the key biological process parameters of sludge yield Y and the food:micro-organism (F:M) ratio. The sludge yield is the amount of biomass generated compared to the amount of organic carbon food source (or substrate) fed into the process. The F:M ratio indicates the availability of food for the biomass, and takes units of inverse time (d^{-1}):

$$F:M = \frac{Q.S}{\mathrm{MLVSS}.V} \tag{2}$$

where Q is the flow rate, S the substrate concentration (or carbon food source – normally expressed as the biochemical or chemical oxygen demand, BOD or COD), and V the tank volume. The MLVSS is the volatile fraction of the mixed liquor suspended solids, and is assumed to equate to the micro-organism concentration (i.e. the biomass). The proportion of MLVSS to MLSS tends to decrease with increasing concentration, but is generally taken as being between 0.75 and 0.85. Increasing the MLSS at a constant feed BOD or COD concentration decreases the *F:M* ratio and, consequently, the sludge production – since the micro-organisms consume more of the organic matter which would otherwise form waste sludge.

Maintenance of the mechanical components of the biological process is largely limited to the FBD aerators, which tend to foul with age in the same way as membranes and, as with membranes, fouling can be both organic and inorganic in nature. Similar cleaning reagents to those used for membranes can thus be used for the aerators, although the aerators generally take longer to foul and are more robust such that higher-strength chemical reagents can be used for cleaning. Alternatively, if they are the inflated rubber sleeve type, the rubber sleeve can be periodically replaced.

2.5 Ancillary equipment

Two key pieces of equipment required for an MBR which exceed that demanded for classical activated sludge treatment comprise (i) fine screening, and (ii) membrane process operation control.

Whilst an MBR can effectively displace primary sedimentation, biotreatment and secondary solid–liquid separation, as well as tertiary effluent polishing, classical bar screens of around 6 mm rating are insufficient for an MBR. Such relatively coarse screens increase the risk of clogging of both membrane channels and, for immersed systems, the coarse bubble aerators. Screen ratings for MBRs range from 3 mm for a FS membrane down to <1 mm for HF configurations. Both the rating and the aperture shape is critical, with circular apertures ("microscreens") providing more rigorous pre-treatment than slit-like apertures ("bar" or "wedgewire" screens). The quantity of screenings generated in an MBR process is therefore considerably greater than that produced by conventional wastewater treatment, such that automated cleaning of the screens is essential for most installations.

Process control is fundamental to all wastewater treatment processes. The design approach for MBR control systems is guided by the application, and in particular the scale of operation, as well as the membrane configuration. Large-scale commercial systems require stability of operation, optimum use of energy and chemicals (specifically for membrane cleaning), and robust control when challenged with changes in hydraulic and/or organic loadings. Perhaps most critically, operation outside of the envelope (specifically permeability and cleaning, Section 2.4.1) stipulated by the membrane technology supplier may invalidate the warranty.

2.6 Applications

It is generally recognised that MBRs become an attractive treatment option when (a) space is limited, and/or (b) a high treated water quality is required (in particular for water reuse). In the past the relatively high cost of MBRs compared with the classical ASP meant that the technology tended to be employed for relatively small flows. However, the increasing stringency of environmental legislation coupled with decreasing capital (specifically membrane) and operating (primarily energy) costs of the process have meant that MBR installations have steadily increased in size in the first 20 years of their implementation, such that there are now a number of municipal MBR plants over 100 megalitres/day (MLD) in peak daily flow (PDF) capacity (Table 2-1) with more under construction or at the planning stage. For example work has begun on Canton WwTP in Ohio (Ovivo) at 333 MLD PDF, and GE's Seine Aval plant (357 MLD PDF) in France is due to come online in 2016.

Table 2-1 World's largest commissioned MBRs (as of June 2014)

Installation	*Supplier*	*Date*	*PDF*	*ADF*
Brightwater, WA	GE	2011	170	117
Qinghe, China	OW/MRC	2011	150	150
North Las Vegas, NV	GE	2011	133	95
Yellow River, GA	GE	2011	111	69
Shiyan Shendinghe, China	OW/MRC	2009	110	110
Aquaviva, Cannes, France	GE	2012	106	59
Busan City, Korea	GE	2012	100	100
Guangzhou, China	Memstar	2010	100	-
Wenyuhe, Beijing, China	OW/Asahi Kasei	2007	100	100
Johns Creek, GA	GE	2009	94	42
Awaza, Turkmenistan	GE	2011	87	69
Jordan Basin WRF, UT	GE	2012	79	53
Beixiaohe, China	Siemens (now Evoqua)	2008	78	-
Al Ansab, Muscat, Oman	Kubota	2010	77	55
Cleveland Bay, Australia	GE	2009	75	29
Broad Run WRF, VA	GE	2008	71	38
Christies Beach, Australia	GE	2011	68	27
Gongchan, Korea	Econity	2012	65	-
Goodna, Australia	GE	2012	64	-
Lusail, Qatar	GE	2011	61	61

PDF Peak daily flow, ADF Average daily flow, GE GE Water and Process Technologies, OW Origin Water, MRC Mitsubishi Rayon Corporation

MBRs have been implemented across a number of industrial sectors. Wastewaters most conducive to treatment by MBR technology are, unsurprisingly, those with a readily biodegradable organic carbon content. Thus the food and beverage sector (Section 3.2) has seen extensive take up of MBR technologies – including both sidestream and immersed configurations. However, wastewaters which have a significant biorefractory content (i.e. sparingly biodegradable) have also been successfully targeted for treatment by MBRs since the long SRTs attainable allow more effective biological treatment than that achieved by conventional biological processes. These include pharmaceutical effluents (Section 3.4) and landfill leachate (Section 3.7). On the other hand, waters containing free (i.e. suspended) oil – either vegetable or mineral in derivation – are somewhat more challenging since membranes are irreversibly fouled by such materials. In such cases pre-treatment – either by plate separation, dissolved air flotation (DAF) or both – is required to protect the membrane. Such unit processes are commonly installed upstream of MBRs for oil and gas wastewaters (Section 3.3) and some food effluent treatment applications (Section 3.2).

2.7 Optimisation and costs

Optimisation of an MBR can be considered from a number of different perspectives:

a) costs: capital expenditure (CAPEX), operating expenditure (OPEX) and total expenditure (TOTEX), where TOTEX is some function of CAPEX and OPEX;
b) carbon footprint, which is normally a function of the energy demand, material usage and waste generated;
c) size, or physical footprint;
d) robustness, as determined by risk of process failure and the requirement for unscheduled human intervention;
e) residuals (i.e. waste streams) and environmental impact, which forms part of the carbon footprint.

The trade-off between capital and operating costs is well recognised. Often, greater investment in capital equipment can increase the process robustness (for example through control measures) and thus decrease the maintenance and repair component of the OPEX. In the case of a membrane process, this is demonstrated by the selection of the design flux. Higher fluxes permit reduced capital costs through the correspondingly reduced membrane area requirement. However, this then increases the likelihood of fouling, demanding more frequent cleaning and increasing the process downtime. The process can also be designed to generate less waste, by extending the SRT and permit extended biodegradation of the organic solids. This then naturally leads to greater energy demand since the higher solids concentration causes (a) a decrease in the oxygen mass transfer rate (Fig. 2-7), and (b) increased clogging propensity of the membrane channels, demanding lower fluxes (Fig. 2-11).

CAPEX and OPEX can be combined to produce total cost provided valid assumptions can be made about plant life, the plant residual value at the end of its life, and trends in value generally (inflation, depreciation, etc). For industrial effluent treatment the primary driver is almost always cost, with the key parameter being the payback period (or return on investment). In many industrial sectors a payback period of more than ~18 months is considered unacceptable, which is significantly constraining unless other mitigating factors exist (such as subsidies or incentives). It is often the case that advanced effluent treatment is demanded to comply with regulations, such as challenging discharge permits or consents, with very onerous punitive measures enforced if compliance is not met.

2.7.1 Capital expenditure (CAPEX)

CAPEX comprises the equipment and installation costs, along with other elements which may or may not be included (the land on which the plant is installed, for example). The basic components of an MBR plant (Fig. 2-12) comprise:

i) *Pre-treatment.* This may include buffering, depending on the temporal variation in the contaminants concentration in the feedwater, and/or sedimentation, if the suspended solids load is high or problematic in some way. Flotation, most typically DAF, is required if the feedwater contains high levels of suspended (or "free") oil or other flotable matter. Fine screening down to 0.8-3 mm is always required, the precise rating depending mainly on the membrane configuration (Section 2.5). The fine screen may be protected by a coarser screen upstream, depending on the gross solids content of the feedwater. A feed pump is always required, the rating depending on the hydrostatic head and flow capacity.

ii) *Process tank.* This is fitted with FBDs connected to the appropriate number and rating of air blowers for the aeration rate required to maintain the target DO level (normally between 0.5 and 2 mg/L). There are normally duty and standby blowers. The tank is usually fitted with agitators for mixing.

iii) *Membrane skid or membrane tank.* A membrane tank is required for immersed membrane technologies. It is normal in an immersed process for the mixed liquor to be pumped at low pressure from the biological tank to the membrane tank at 3-5 times the feed flow rate. The overflow (the unpermeated or retentate flow) is then directed either back to the aeration tank (Fig. 2-1a) or to an anoxic zone (Figs. 2-5, 2-6). For a sidestream technology the same biological process configurations apply, but in this case the membranes are mounted in a skid. The flow rate through the skid is partly determined by the CFV requirement (the flow divided by the total membrane channel cross-sectional area).

iv) *Membrane modules.* For immersed technologies blowers are required (normally duty/standby, as with the process air blowers) for membrane air scouring via coarse bubble diffusers (CBDs). These may be either fitted on the membrane frame or integrated with the membrane module itself (Fig. 2-8). HF iMBRs are normally fitted with permeate suction pumps. FS iMBRs may also have permeate pumps or may operate simply by the hydrostatic head provided by the height of the sludge above the membrane outlet. Permeate pumps must be self-priming.

All sMBRs operate by a positive pressure of 2-4 bar on the retentate side, with an external sludge pump providing the necessary CFV of 1.5-3.5 m/s. For a classic pumped system the membranes are linked to form a serpentine path (Fig. 2-3c).

v) *Chemical cleaning equipment*. This will normally comprise low flow/low pressure pumps with appropriate storage for the hypochlorite and mineral/organic acid chemicals.

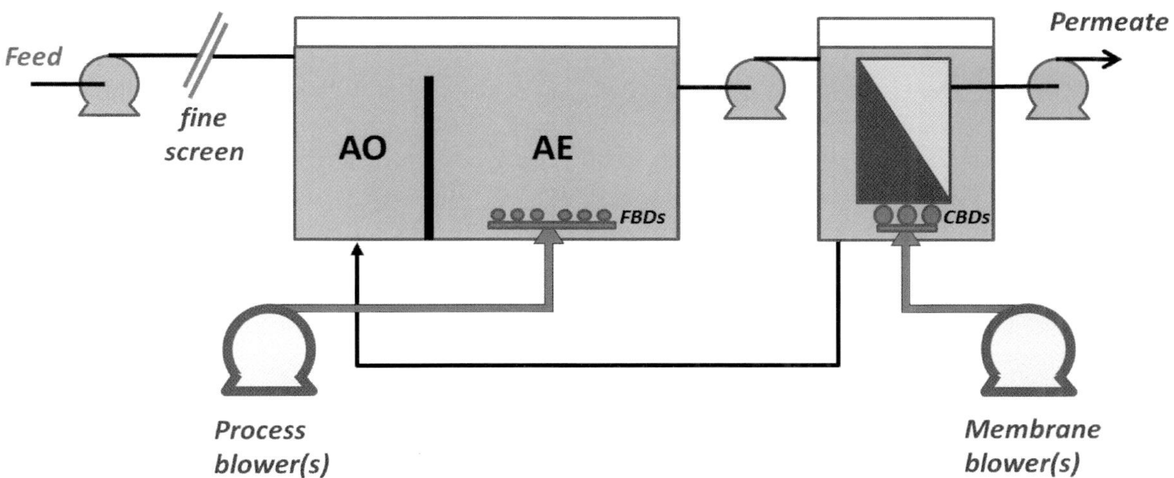

Figure 2-12 Basic process flow diagram (AO: anoxic; AE: aerobic).

vi) *Process control equipment*. Process control is more complex than is required for a CAS process since it encompasses both the biological process side and the membrane maintenance (physical and chemical cleaning).

vii) *Supplementary equipment and/or unit operations*. These may include sludge dewatering and associated pumping, biological and/or chemical P removal (demanding further sludge transfer and tankage, and/or chemical dosing for phosphate precipitation), permeate disinfection (chlorine dosing or UV irradiation) or various other additional processes for the main process or waste streams.

The total cost of the MBR therefore relates to:
(a) the cost of the individual components (i)-(vii) above,
(b) the operational costs associated with these components,
(c) supplementary site-specific costs (e.g. waste disposal, land, installation, etc).

Component costs are not readily quantifiable, since they depend primarily on the flow capacity of the installation, but also on the plant technical specification and geographical region. The depiction of cost as a function of flow rate is often termed a "cost curve", and there is nearly always a degree of economy of scale: costs per m^3 of treated water decrease with increasing plant size. The economy of scale is generally less for an MBR than for a CAS plant, however, since the unit membrane area cost generally varies little with the amount required. This is to be distinguished from, say, a concrete tank where the cost per unit volume of water treated (i.e. the specific cost) decreases with increasing size of the tank. Whilst the precise relationship between CAPEX and plant flow capacity – and so the extent of the economy of scale – varies considerably and has to be considered on a case-by-case basis, the OPEX is more readily estimated.

2.7.2 Operating expenditure (OPEX)

The key components of OPEX (ignoring labour) are the energy demand, critical component replacement (primarily the membrane itself), waste management and chemicals costs. Energy costs relate to air and liquid pumping, with aeration energy always being the dominant energy demand constituent. Critical component replacement costs relate mainly to membrane replacement, and waste management costs are very site-specific. For most small-to-medium industrial installations where it is less likely that extensive sludge management technologies (such as anaerobic digestion, AD) exist on site, the investment cost of a sludge dewatering process may be offset by the reduced sludge tankering costs. All other OPEX contributions, including chemicals costs, tend to be very minor.

2.7.2.1 Energy

Energy demand primarily relates to aeration (for process and membrane) and sludge transfers, with permeate production also demanding energy to differing extents depending on the configuration of the MBR process. Its determination from first principles is more challenging. Process-related equations from mathematical models dating back several decades (Ekama et al., 1984) and derived from the CAS process have been directly transferred to MBR biological processes. Such expressions allow aeration energy demand, along with other key process parameters such as tank sizes and sludge discharge rates, to be calculated as a function of feed and effluent concentrations BOD, COD and TKN concentrations (Judd and Judd, 2011), with some approximations provided to allow very simple analytical expressions (Yoon, 2011). There is some debate as to whether the fundamental constants originally determined for the activated sludge models describing CAS are equally applicable to MBRs, since the sludge for the latter appears to differ physicochemically from that of CAS processes. However, these models have nonetheless reached a high level of sophistication, enabling the modelling of the dynamic behaviour of the biotreatment process over wide-ranging conditions. Most recent developments have encompassed organic foulant generation (soluble microbial product, SMP, and extracellular polymeric substances, EPS).

Biochemical oxygen consumption estimation

Whilst the computer models allow extensive study of the process virtually, the oxygen demand for biological treatment can be calculated analytically from a simple mass balance (Tchobanoglous et al., 2013). Under steady-state conditions (i.e. constant MLSS concentration), the oxygen requirement relates to the difference between the mass flow of substrate consumed and the proportion of mass flow of the wasted sludge which contains the substrate. The mass flow of substrate consumed is simply $Q_F\Delta S$, where Q_F is the feed flow and ΔS the difference between the feed and permeate COD or TKN substrate concentration. The proportion of the mass flow of substrate in the wasted sludge flowing at a rate of Q_W is $\lambda Q_W X$, where λ is the ratio of substrate to MLSS concentration and the latter is denoted X.

Since, in the case of the organic carbon substrate, the concentration is defined directly as oxygen demand in the form of COD, the oxygen consumption rate in mass per unit time is given by:

$$O_{2,COD} = Q_F\Delta S_{COD} - \lambda_{COD} Q_W X \quad (3)$$

where λ_{COD} is generally taken as being ~1.1 kg COD per kg MLSS (Yoon, 2011).

Q_W is normally expressed as a fraction of Q_F by the observed sludge yield Y_{obs} in kg biomass per kg COD, which is generally slightly lower than the theoretical yield Y (Section 2.4.2) and is dependent on the SRT and the nature of the organic matter:

$$Y_{obs} = Q_W X/(Q_F\Delta S_{COD}) \quad (4)$$

So Equation (3) becomes:

$$O_{2,COD} = Q_F \Delta S_{COD} (1 - \lambda_{COD} Y_{obs}) \quad (5)$$

For nitrogen as TKN, the concentration must be adjusted for the biochemical reaction stoichiometry, i.e. the ratio of the oxygen to substrate in the biochemical reaction. For a 2:1 O_2:N reaction stoichiometry the O:N mass ratio is 4.57 (Section 2.3.5.1):

$$O_{2,TKN} = 4.57 Q_F (\Delta S_{TKN} - \lambda_{TKN} Y_{obs} \Delta S_{COD}) \quad (6)$$

where λ_{TKN} is the biomass TKN content in TKN per g MLSS and is 0.095, based on a λ_{COD} value of 1.1 (Yoon, 2011).

If denitrification is included through the MLE process (Figure 2-5) then the oxygen demand is reduced slightly since the nitrate ion (NO_3^-), through its reduction to nitrogen gas (N_2), provides the equivalent of 1.25 molecules of oxygen. This equates to a mass ratio of 2.86 O:N, and the equivalent oxygen benefit is then:

$$O_{2,Denit} = -2.86\, Q_F (\Delta S_{TKN} - \lambda_{TKN} Y_{obs} \Delta S_{COD} + \Delta S_{Nitrate}) \quad (7)$$

The overall oxygen concentration demand in mass per unit volume of water treated (g/m^3) is obtained by adding the three contributions together and dividing by the flow Q_F:

$$D_{O2} = \Delta S_{COD} (1 - \lambda_{COD} Y_{obs} - 1.71 \lambda_{TKN} Y_{obs}) + 1.71 \Delta S_{TKN} - 2.86 \Delta S_{Nitrate} \quad (8)$$

Aeration rate

The specific process biological aeration demand in Nm^3/m^3 can be estimated from the above oxygen consumption rate through the use of empirical parameters which relate to the mass transfer of oxygen from air into the mixed liquor:

$$Q_A/Q_F = SAD_{bio} = D_{O2}/(\rho_A C'_A\, SOTE\, y\, \alpha\, \beta\, \gamma) \quad (9)$$

In the above equation ρ_A and C'_A refer to the air density and the fractional oxygen concentration (~0.21) respectively. *SOTE* is the standard oxygen transfer efficiency in terms of the percentage of oxygen transferred into the liquid phase per m depth of pure water, and y is the depth of the aerator in the tank. The terms α, β, and γ are OTE correction factors for solids, salinity and temperature. Of these, the α factor is the most significant (Section 2.3.5.1), with a number of different correlations reported with MLSS concentration (Figure 2-7). The relationship is complicated by other interdependencies, in particular the variation of *SOTE* with flow rate. Finally, aeration of the process tank must also normally provide sufficient agitation for mixing, which effectively places a lower limit both on the aeration rate and the bubble size.

The specific aeration energy demand for air pumping in kWh per Nm^3 air is a function of inlet and outlet pressure ($P_{A,in}$ and $P_{A,out}$ respectively, where the outlet pressure is approximately given by the inlet plus the pressure exerted by the head of water above the aerator) and blower efficiency ε. It is given by expressions such as:

$$E'_A = k\, P_{A,in} \left(\left(\frac{P_{A,out}}{P_{A,in}} \right)^{0.283} - 1 \right) \quad (10)$$

where k is a dimensionless function of temperature and blower efficiency but is generally between 6 and 7.5 for pressure quoted in SI units (i.e. Pascals). The product of E'_A and Q_A in Nm^3/h yields the required power rating of the blower in kW. The product of E'_A and SAD_{bio} provides the specific energy demand for biological aeration $E_{L,bio}$.

Contrary to the advanced state of the modelling of biochemical oxygen demand, there is currently no agreed method for the calculation of iMBR membrane air scour rates. Values adopted for these are generally those recommended by the supplier. Further elucidation of energy demand is therefore largely derived from a combination of empirical data, such as pump curves for liquid pumping, and heuristic expressions.

Heuristic expressions

Heuristic determination of energy demand for an existing installation is simply through the division of the equipment power rating in kW by the permeate flow in m^3/h to give the specific energy demand, or *SED* (or simply E), in kWh/m^3. In the case of membrane aeration, the energy demand is given by SAD_p multiplied by E'_A. The energy demand component of the OPEX is given by the addition of all the power ratings, divided by the installed flow capacity:

$$E_{total} = \Sigma E_L + SAD_p\, E'_{A,m} + SAD_{bio}\, E'_{A,bio} + \Sigma E_{other} \qquad (11)$$

where E_L relates to liquid pumping operations, i.e. for sludge transfer ($E_{L,sludge}$), permeation through the membrane ($E_{L,m}$), and cleaning chemicals ($E_{L,chem}$), SAD_p (or SAD_m/J) and SAD_{bio} (Equation (9)) are the specific aeration demands in Nm^3 air per m^3 permeate for membrane scouring and process biological aeration respectively, and E_{other} encompasses energy demanded for mixing (E_{mix}) and process control (E_{contr}). Mechanical mixing of the anoxic and anaerobic zones is required since there is no aeration in these regions to provide reactor mixing. Mixing of equalisation (EQ) tanks may also be conducted mechanically. Energy consumption for process control generally makes up less than 2% of the total energy demand.

Analytical equations defining the relationship between blower power and air flow rate and, in particular, sludge pump power and sludge flow rate can be complex if all variables are included. The key variables contributing to energy demand for both membrane and biological aeration are the depth of the aerator in the tank and the type of aerator (Section 2.3.5). In the case of liquid pumping the main variables are the resistance to flow (introducing pressure losses in the system) and the liquid viscosity, but in practice the former has far greater impact because resistances increase during periods of low flow making pumping less efficient overall.

As a rule-of-thumb $E'_{A,m}$ is normally around 4.5-6.5 Wh per Nm^3 of air per m of depth of aerator in the tank for both membrane and biological aeration. For sludge transfer operations the energy demanded per unit volume of sludge or permeate pumped depends on a number of factors relating to the process design. These include the length and diameter of the pipework (which directly impact on the pressure drop along it), the velocity in the pipes and membrane channels (for sMBRs), and the TMP. Since pressure losses increase with increased velocity, the E_L is always higher for pumped sMBRs where a high CFV is required, as well as a relatively high TMP. For sludge transfer operations at a constant rate and with negligible pressure difference between the inlet and outlet the energy demand is normally between 15 and 20 Wh per m^3 of sludge or water pumped.

For a membrane aerator immersion depth of 2.5 to 5 m, E'_A and E_L are comparable in value. However, since E'_A is multiplied by SAD_p to obtain the energy demand for membrane aeration E'_A in kWh/m^3 permeate, the energy demand for air scouring of an immersed membrane is always higher than that demanded for liquid transfers. Since the energy demand is inversely proportional to the overall efficiency of the pumps and blowers, i.e. the efficiency with which the electrical energy input is converted into operational power, then factors which adversely affect efficiency increase the energy demanded. This becomes important if the pumps and blowers are required to be turned down during periods of low flow, since this lowers their operational efficiency.

sMBR pumping energy demand

The specific energy demand for a sidestream process where a liquid is being pumped along a channel under pressure is derived from Bernoulli's equation:

$$SED_m = [H + \Delta P/(\rho g) + v^2/(2g)]\,\rho g/(\varepsilon_{tot}\,\theta) \quad (12)$$

where H is the "static head" in m water (height of water above the discharge point), $\Delta P/(\rho g)$ is the "pressure head" from the applied pressure, and $v^2/(2g)$ the "velocity head", v being the crossflow velocity, ρ the liquid density and g the acceleration due to gravity. ε_{tot} and θ are respectively the sum of the efficiencies (motor, pump, gear box) and the overall conversion – the proportion of the liquid pumped through the membrane channels that is converted to permeate.

ΔP can be determined from engineering hydraulics but is normally in the region of 3-4 bar (300,000 – 400,000 Pa) for a 2-4 loop sidestream such as that shown in Fig. 2-3(c). Given that the density is ~1,000 kg/m^3 and v is normally less than 5 m/s, it follows that the velocity head is negligible compared with the sum of the pressure head and the static head. An example calculation based on Equation 12 and an MT membrane module typically employed for sMBR technologies is provided in Annex 1.

2.7.2.2 Other operating cost factors

Energy demand represents the most significant contribution to operating cost. In most cases the second biggest contributor to OPEX is critical component replacement which, in the case of MBRs, is largely associated with the membranes themselves. For practical purposes, the membrane life (and thus replacement frequency) can be assumed to be determined by the warranty. Reported membrane replacement frequency information is scarce, but published papers in this area (Ayala et al., 2011; Côté et al., 2012) suggest that a life of 8-10 years for a modern membrane applied to municipal wastewater treatment has been satisfactorily demonstrated. No corresponding research appears to have been published assessing membrane life for industrial applications; the most recently-produced robust membranes have not been running long enough to corroborate life expectancy. It may however be expected for membrane life to be adversely affected by more aggressive operating conditions and chemical cleaning protocols, such as may arise for industrial effluent treatment, and anecdotal evidence suggests that suppliers' warranties are shorter for some industrial applications than for municipal ones. Replacement of other critical components (e.g. aerators or aerator sleeves, valves or valve seats, permeate tubing for FS membrane panels) does not generally add significantly to costs.

An OPEX item which varies significantly according to the specifics of the installation is the sludge disposal costs. The sludge generation can be calculated from the same biochemical approach as used to determine the process performance, and is also a function of the loading (i.e. the total mass of COD or BOD per unit time). In general the yield of sludge Y (Section 2.4.2) is within the range 0.4-0.6 g volatile suspended solids (VSS) per g COD, but varies according to the SRT and the nature of the organic matter. Unless an AD facility exists on site, or some other disposal route of marginal cost, it is normally desirable to thicken the sludge (i.e. remove water) prior to its transport off site to a centralised sludge management facility.

The extent of thickening or dewatering attainable (i.e. sludge volume reduction by reducing the water content) depends on the technology used. Technologies include simple sedimentation, gravity or pressurised belt filters, filter bags, filter presses and centrifuges, with the degree of water removal generally increasing with the energy input. There also exist membrane thickening technologies, generally capable of achieving 4-4.5 wt% solids concentration, which are based on the same membrane technology as that used for the main MBR process. The requirement for thickening can contribute as much to the cost as the membrane aeration.

A further minor OPEX contributor is the chemical demand for membrane cleaning. This will naturally fluctuate with the cost of the chemicals, and specifically sodium hypochlorite and citric acid for the vast majority of applications, but rarely contributes more than 2-3% of the total OPEX.

Finally, labour costs vary considerably globally and also between installations, since plants may be designed with different degrees of manual intervention envisaged. Since events leading to unscheduled manual intervention are the main cause for concern and supplementary costs with respect to labour, it is inevitable that costs associated with such events will form a larger proportion of the total OPEX for smaller plants than for larger ones were the events to occur at the same frequency. Small plants are therefore usually designed more conservatively than larger ones, especially given that peak loading factors (i.e. the ratio of peak to average flow) also tend to increase with decreasing plant size.

2.7.2.3 Total OPEX

An estimate of the OPEX from basic design parameter values is given by:

$$\text{OPEX} = L_E(E'_{A,m}SAD_m/J + E_{L,sludge}R + E_{L,m} + E'_{A,bio}SAD_{bio}) + L_M/(J\,t) + L_C + L_W + L_L \qquad (13)$$

where L_E is the cost of electrical energy in \$/kWh, $E_{L,sludge}$ and $E_{L,m}$ are the respective energy demands in kWh/m^3 permeate generated for sludge pumping and permeate extraction, R is the recycle ratio (recycle flow per feed flow), L_M is the membrane cost per m^2 membrane area, J is the flux in appropriate units and t the membrane life in days. L_C, L_W and L_L represent the specific costs per m^3 treated water for chemicals consumption, waste disposal and labour. Whereas L_C can be accurately determined from chemicals usage arising from the cleaning protocols, labour costs are a function of the extent of manual intervention required and rely on historical records for their estimation. All other terms are as given in Equation (11).

For an immersed system $E_{L,m}$ is a function of the transmembrane pressure (TMP). For gravity-driven flat sheet technologies, which can operate at hydrostatic heads of as low as 0.08 m, the contribution to energy demand is negligible. For pumped sidestream systems, for which E'_A is zero, the permeation energy represents a significant proportion of the total energy demand – perhaps as high as 2 kWh/m^3 depending on the flux attained for the energy applied.

The energy demanded for biotreatment ($E_{L,bio} = E'_{A,bio}SAD_{bio}$) is a function of the biochemical oxygen consumption (Section 2.7.2.1), which increases with organic loading (as COD and TKN) and MLSS concentration (Section 2.3.4). Calculation of the specific biological process aeration demand proceeds via Equation (8), which determines the oxygen demand, and Equation (9), which converts the oxygen demand to aeration demand based on the operating conditions and the type of aerator employed. This is converted to energy demand by multiplying by the specific aeration energy demand for air pumping E'_A in kWh per Nm^3 air, Equation (10), to provide $E_{L,bio}$ in kWh/m^3 permeate.

For a given set of operating conditions, the relationship between biological energy demand (incurred by process aeration) and organic removal rate is approximately linear. For example, based on a feedwater TKN concentration as N of around 5% of the COD concentration, a target effluent ammonia of 0.5 mg/L and a biotank MLSS of 8 g/L, then $E_{L,bio}$ in kWh per m^3 treated water is around half of the feed COD concentration in g/L. For membrane aeration, SAD_m values are recommended by technology suppliers and may be adjusted for specific applications and/or flow conditions. Values generally range between 0.2 and 0.55 Nm^3/(m^2.h), with the smallest values tending to be associated with HF modules and larger ones with FS technologies.

An example calculation of the biological energy component is appended (Annex 1) based on the conditions listed in Table 2-2. A correlation of total OPEX with two of the most influential cost components, membrane life and net flux (Fig. 2-13), based on the listed parameter values and Equation (13), indicates the sensitivity of OPEX to these two parameters. The doubling of the membrane life decreases the OPEX by 20-30% in this case, whereas doubling the flux decreases it by 30-38%. There is obviously greater sensitivity to membrane life at lower fluxes since the amount of water treated over the same time period decreases. Finally, sensitivity to membrane operational factors, and in particular SAD_p, increases with decreasing organic loading since the membrane aeration makes up a greater proportion of the total energy demand if the process aeration is decreased.

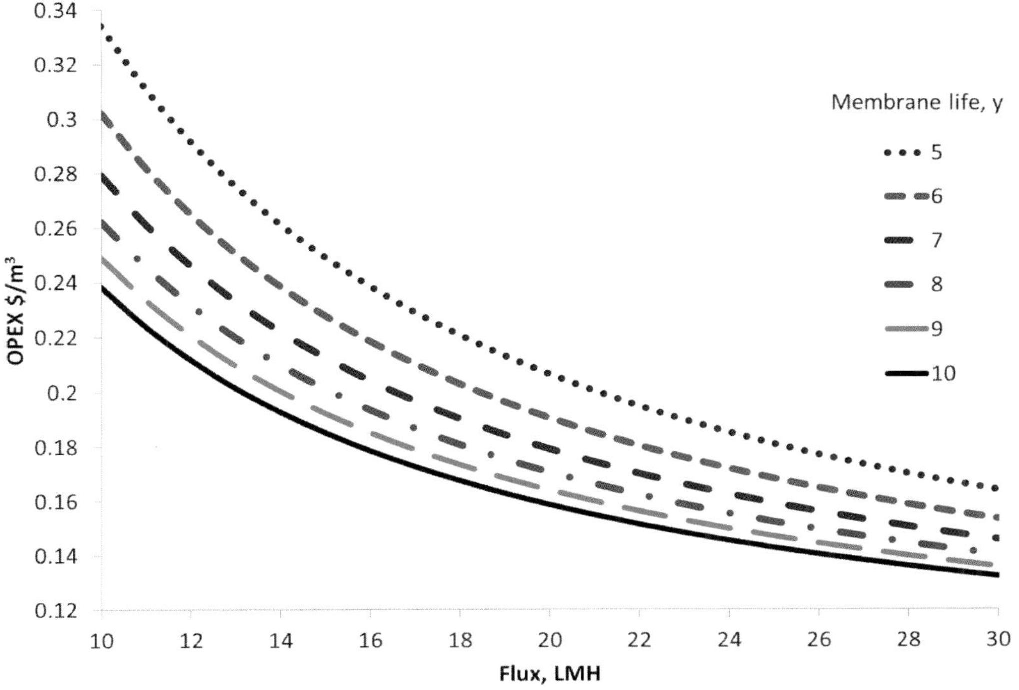

Figure 2-13 OPEX as a function of net flux and membrane life (excl. labour and waste disposal costs)

Table 2-2 Base parameter values for OPEX determination

Parameter	Symbol	Units	Value
COD removed	ΔS_{COD}	mg/L	1,000
TKN removed	ΔS_{TKN}	mg/L	50
Specific aeration demand against membrane area[1]	SAD_m	$Nm^3/(m^2.h)$	0.25
Specific energy demand for permeation	$E_{L,m}$	kWh/m^3	0.015
Energy demand of sludge pumping (power/sludge flow)	$E_{L,sludge}$	kWh/m^3	0.02
Specific energy demand of process aeration[2]	$E_{L,bio}$	kWh/m^3	0.5
Energy demand of blower (power/air flow delivered)[2]	E'_A	kWh/Nm^3	0.021
Recycle ratio: membrane-biotank + between biotanks	R	-	4
Electrical energy cost	L_E	\$/kWh	0.12
Membrane cost	L_M	$\$/m^2$	80
Chemicals cost, 10-15% hypo and 50% citric[3]	L_C	$\$/m^3$	0.008

[1] *Usually recommended by the technology supplier*
[2] *See Annex 1 for full calculation*
[3] *Reagents priced at \$1,800/te 10-15% hypo and \$860/te citric*

2.7.3 Total cost

Total expenditure (TOTEX), the combination of CAPEX and OPEX, can be represented by the net present value (NPV):

$$\mathrm{NPV} = \sum_{t=0}^{t} \frac{\mathrm{CAPEX}_t + \mathrm{OPEX}_t}{(1+i)^t} \tag{14}$$

where t is the plant age and i the discount rate, which is normally taken as the rate of return on the investment were the capital sum to be invested on the financial markets.

The calculation proceeds by determining the expenditure for each year between the year of installation and the end of plant life t, after which time its residual value has to be estimated (normally assumed to be negligible in comparison with all other expenditure). The sum of the total outgoings and income over the entire life of the plant then represent the NPV, and account for the fact that certain expenditure items (such as membrane replacement) occur at junctures when the value of the money itself differs from that at the start or end of the plant life. The NPV is thus a more accurate representation of the actual cost than the simpler definition of OPEX provided in Equation 14 since it accounts for this change in the value of the money through the inclusion of i.

Determination of NPV becomes especially important in cost benefit analyses, where income might be associated with a real or notional value of the product water. NPV analyses almost invariably reveal the benefit of more conservative design, incurring a higher CAPEX, in mitigating OPEX and thus TOTEX. This is a natural outcome of the far greater contribution of OPEX to TOTEX compared with the initial capital outlay for a reasonable plant life. Thus, for example, the installation of a buffering tank to reduce peak loading and thus the membrane area requirement (and with it the total membrane scouring aeration demand) has been shown to significantly reduce the overall TOTEX over the course of the 25-year plant life (Verrecht et al., 2010).

References

Ayala, D., Ferre, V., and Judd, S.J. (2011). Membrane life determination in full-scale immersed membrane bioreactors, *J. Membrane Sci.*, **378** 95–100.

BCC (2008). *Membrane Bioreactors: Global Markets*, BCC Report MST047B.

Christian, S., Grant, S., Wilson, D., McCarthy, P., Mills, D., and Kolakowski, M (2010). The first two years of full-scale anaerobic membrane bioreactor (AnMBR) operation treating high-strength industrial wastewater. *WEFTEC 2010*, 2-6 Oct 2010, New Orleans.

Cornel, P., Wagner, M. & Krause, S. (2003). Investigation of oxygen transfer rates in full scale membrane bioreactors. *Water Sci. Technol.* **47**(11), 313-319.

Côté, P., Alam, Z. and Penny, J. (2012). Hollow fiber membrane life in membrane bioreactors (MBR), *Desalination* **288** 145-151.

De la Torre, T., Rodŕiguez, C., Gómez, M.A., Alonso, E., and Malfeito J.J. (2013). The IFAS-MBR process: A compact combination of biofilm and MBR technology as RO pre-treatment. *Desalination and Water Treatment* **51**(4-6) 1063-1069.

Dereli, R.K., Ersahin, M.E., Ozgun, H., Ozturk, I., Jeison, D., and van der Zee, F., van Lier, J.B. (2012). Potentials of anaerobic membrane bioreactors to overcome treatment limitations induced by industrial wastewaters. *Bioresource Technol.* **122** 160-170.

Diez, V., Ramos, C., Cabezas, J.L., (2012). Treating wastewater with high oil and grease content using an Anaerobic Membrane Bioreactor (AnMBR). Filtration and cleaning assays. *Water Sci. Technol.* **65** (10), 1847–1853.

Ekama, G.A., Marais, G.v.R., Siebritz, I.P., Pitman, A.R., Keay, G.F.P., Buchan, L., Gerber, A., Smollen, M., (1984). Theory Design and Operation of Nutrient Removal Activated Sludge Processes. Water Research Commission, Pretoria, South Africa, ISBN 0 908356 13 7.

Frost and Sullivan (2013). *Global Membrane Bioreactor (MBR) Market*, 11 Jan 2013.

Germain, E., Nelles, F., Drews, A., Pearce, P., Kraume, M., Reid, E., Judd, S. J. & Stephenson, T. (2007). Biomass effects on oxygen transfer in membrane bioreactors. *Water Res.* **41** (5) 1038–1044.

Henkel, J., Cornel, P., and Wagner, M. (2011). Oxygen transfer in activated sludge – new insights and potentials for cost saving. *Water Sci. Technol.* **63**(12) 3034-3038.

Judd, S., and Judd, C. (2011). The MBR Book, 2nd Edition, Butterworth-Heinemann (Oxford, UK).

Kayser, R. (1967). *Ermittlung der Sauerstoffzufuhr von Abwasserbelüftern unter Betriebsbedingungen*. Dissertation, Technische Hochschule Braunschweig, Braunschweig.

Krampe, J. (2001). *Das SBR-Membranbelebungsverfahren*. Dissertation, Universität Stuttgart, Stuttgart.

MarketsandMarkets (2013), Membrane Bioreactor Systems Market by Types, Configuration & Applications - Trends & Forecasts To 2017, marketsandmarkets

Muller, E. B., Stouthamer, A. H., van Verseveld, H. W. and Eikelboom, D. H. (1995). Aerobic domestic waste water treatment in a pilot plant with complete sludge retention by cross-flow filtration. *Water Res* **29**(4) 1179–1189.

Ødegaard, H., Mende, U., Skjerping, E.O., Simonsen, S., Strube, R., Bundgaard, E. (2012). Compact tertiary treatment based on the combination of MBBR and contained hollow fibre UF-membranes, *Desalination Water Treat.* **42**(1-3) 80-86.

Robles, A., Durán, F., Ruano, M.V., Ribes, J., and Ferrer, J. (2012). Influence of total solids concentration on membrane permeability in a submerged hollow-fibre anaerobic membrane bioreactor. *Water Science and Technology*, **66**(2), 377-383 (2012).

Rosenberger, S. (2003). Charakterisierung von belebtem Schlamm in Membranbelebungsreaktoren zur Abwasserreinigung. Dissertation, VDI Verlag GmbH, Düsseldorf.

Santos, A., Ma., W and Judd, S. J. (2011). Membrane bioreactor technology: two decades of research and implementation, *Desalination*, **273**, 148–154.

Scheumann, R. and Kraume, M. (2009). Influence of hydraulic retention time on the operation of a submerged membrane sequencing batch reactor (SM-SBR) for the treatment of greywater. *Desalination* **246**(1-3) 444-451.

Singhania, R.R., Christophe, G., Perchet, G., Troquet, J., and Larroche, C. (2012). Immersed membrane bioreactors: An overview with special emphasis on anaerobic bioprocesses. *Bioresource Technology* **122** 171-180.

Skouteris, G., Hermosilla, D., López, P., Negro, C., and Blanco, Á. (2012). Anaerobic membrane bioreactors for wastewater treatment: A review. *Chemical Engineering Journal* **198-199** 138-148.

Smith, A.L., Stadler, L.B., Love, N.G., Skerlos, S.J., Raskin, L. (2012). Perspectives on anaerobic membrane bioreactor treatment of domestic wastewater: A critical review. *Bioresource Technology* **122** 149-159.

Tchobanoglous, G., Burton, F.L., and Stensel, H.D. (2013). Wastewater engineering: treatment and reuse, 5th ed., McGraw-Hill, New York.

van der Roest, H. F., van Bentem, A. G. N. & Lawrence, D. P. (2002). MBR-technology in municipal wastewater treatment: challenging the traditional treatment technologies. *Water Sci. Technol.* **46** (4-5) 273–280.

Verrecht, B., Nopens, I., Maere, T., Brepols, C., and Judd, S. (2010). The cost of a large-scale hollow fibre MBR, *Water Res.* **44** 5274-5283.

Visvanathan, C., Abeynayaka, A. (2012). Developments and future potentials of anaerobic membrane bioreactors (AnMBRs). *Membrane Water Treatment* **3(1)** 1-23.

Yoon, S.H. (2011). MBR Design, www.onlinembr.info, accessed 03 March 2014.

3 Industrial effluent quality and treatment

3.1 Industrial vs. municipal effluent treatment

A number of general issues arise in industrial effluent treatment distinguishing it from the treatment of municipal wastewater.

Firstly, the original decision to install an industrial wastewater treatment plant (IWwTP) is almost entirely financial. It may be based on long-term strategic planning, be a necessary component of a new-build facility in the absence of a sewer (or else a condition imposed by the sewerage operator) or be an imposed regulatory requirement in the face of onerous financial penalties. Thus, regardless of the precise motivation, the cost analysis – and specifically the determination of the payback period – is crucial.

If a sewage connection exists and the sewerage operator (normally the statutory or private water/wastewater company or operator) is prepared to accept the discharged effluent then this would be expected to be the most cost effective route. Operational costs are then determined by the fee levied by the sewerage operator for receiving the effluent. This is normally on a pollutant load basis, i.e. per kg of COD and/or suspended solids per unit volume or time, and is most likely to be less than the capital cost of a treatment plant if amortised over a period of 18 months or less. 18 months is a reasonably representative period over which most companies would require payback of the capital investment.

It is only possible to discharge to a municipal works via the sewer if (a) there is spare capacity on the works, and (b) the works is capable of treating the industrial effluent without upset to it. Moreover, local or global market forces can impact on any decision made. The sewerage operator may regard the loss of the industrial effluent as being onerous both financially and technically, since there would be both a loss of income and a possible upset to the centralised wastewater treatment works (WwTW) if it becomes under-loaded; an under-loading of a biological treatment process produces a sub-optimal performance. The sewerage operator may then choose to offer discretionary rates to allow the company to continue to discharge to their WwTW. Wider market forces come into play when it comes to operations within the industrial facility itself. The company may choose to cease production of a specific range of goods or commodity item within the facility, which then impacts on the effluent quality. This subsequently affects both the charge levied by the sewerage operator for receiving and treating the effluent and/or the viability of the effluent treatment process – whether this is the IWwTP or the centralised WwTW.

Regardless of any machinations surrounding wastewater discharge levies to sewer, it is normally simpler and less expensive to obtain permission to discharge to the sewer than to an environmental body. If, on the other hand, no sewer connection exists then it is almost inevitable that an IWwTP must be installed. It is then a matter of designing a system which is fit for purpose at the lowest cost, whilst still maintaining the principle of the best available technology.

Industrial wastewater treatment technologies and systems tend to be more diverse than municipal ones, reflecting the much wider range of pollutants, their concentrations and their temporal variability in industrial effluents. Whilst there are usually no impacts from infiltration or hydraulic surges from storm events for an IWwTP, fluctuations in flows and loads are generally greater for industrial effluents than municipal ones. These primarily correspond to addition and losses of materials associated with the manufacturing or other industrial processes within the facility. Such fluctuations demand flow equalisation, which is standard practice at many industrial sites whereas it is normally limited by expense for municipal WwTW. The provision of 12-24h of EQ then allows a degree of process control, if samples are taken from the

EQ tank and appropriate action can be taken for mitigating any potential adverse impacts of recorded changes in water quality.

Such process control measures are of critical importance for treating industrial effluents. The biological unit operation of an IWwTP must be kept viable during seasonal shutdown periods, such as winter vacations, and may also be required to be robust to temperature changes if the latter are not adequately addressed by EQ. Sudden falls in temperature can adversely affect the microbial ecology generally and nitrification specifically, and nitrification is also sensitive to pH decreases to 5.5 or less. In the worst case, reseeding of the bioreactor with new sludge may become necessary, but new sludge is invariably from a municipal source which must then be acclimatised to the industrial effluent.

The sludge product can vary widely in manageability. A food effluent treatment sludge is likely to be more acceptable for composting than municipal sludge, because of reduced perceived pathogen risks. On the other hand, some industrial MBR sludges may be categorised as hazardous wastes due to, for example, their heavy metal content. By the same token, the content of industrial effluent itself (with respect to heavy metals, biorefractory or volatile organic matter, and other substances considered onerous to the treatment process) can cause problems with the sludge management at receiving municipal WwTWs. If the quality of the sludge generated changes to the point that its required end disposal route changes from a relatively low cost option (such as spreading to land) to a high-cost one (incineration or landfill), then the sewerage operator may choose either to impose a higher levy for accepting the effluent or else refuse to accept it on the basis of the presence of the onerous contaminants.

It is perhaps the changeability of the industrial effluent which affords the greatest opportunities for MBRs. Whilst EQ offers the greatest practical contingency against load variations, expansion of operations within an industrial facility resulting in an increase in pollutant load demands that the effluent treatment plant be adapted accordingly. If, as is often the case, spatial restrictions exist which limit the opportunity for additional EQ and/or the expansion of the existing unit processes, then the remaining option is to intensify the treatment capability of the existing assets. A simple calculation based on the relative footprint of a membrane tank and a secondary clarifier for solid liquid separation (Section 4.5) reveals that immersed MBR membrane technologies generally offer a 2-8 times reduction in footprint compared with that of a secondary clarifier. For pumped sidestream technologies, the footprint reduction may be even greater since the upper limit on the skid height is determined only by on-site height constraints rather than the membrane module physical dimensions.

3.2 Food and beverage

The food and beverage (F&B) sector produces both finished products destined for consumption and intermediate products destined for further processing. It is diverse compared to many other industrial sectors, including dairy, maltings, breweries, distilleries, wineries, soft drinks, cereals, potato chips, salads and produce, coffee, confectionery, edible oils, meat and poultry processing and various other prepared foods. A wide range of raw materials, products and processes thus exist with numerous combinations of each in the various markets from specialist regional products to globally-recognised ones. The sector is heavily reliant on good quality water, it being the principal ingredient in all beverages and a number of foods. However, its principal use is for ancillary processes including washing, cooling, heating, cooking, conveying plus cleaning and sanitising of equipment and to provide site utilities.

Effluent generated in the F&B industry is typically high in organic loading, with BOD and COD concentrations that can be 5-100 times higher than for domestic wastewater. The TSS also varies from negligible to very high concentrations, and may contain FOG (fats, oils and grease, e.g. from meat, fish, dairy and vegetable oil production) and high levels of ammonia and/or phosphorus if large quantities of phosphoric acid are used in the process, such as for vegetable oil de-gumming or cleaning operations. Equally, some F&B wastewater can be deficient in N & P and require nutrient balancing. However, they are generally readily biodegradable with COD/BOD ratios ranging from ~0.4-0.5 for bakery products to >0.8 for poultry processing.

Membranes have been used in the F&B industry for nearly 50 years, but mostly for process rather than effluent treatment. The implementation of membranes and in particular MBR technology for wastewater treatment has only occurred since the mid-late 1990s, with market penetration following the same pattern as for municipal wastewater for much the same reasons (declining MBR costs, increasing freshwater supply costs, increasing stringency of legislation, improved technology design and reliability, spatial restrictions, etc). Examples of iFS applications in Japan date back to the mid-1990s. Early examples of sMBR technology applications in Europe include an 800 m^3/d dairy in France, based on ceramic membranes from Degremont, which was commissioned in 1997, along with the Orelis 'plate and frame' MBR for a 50-120 m^3/d dairy. MBR technology applications have now extended across almost all sectors of the F&B industry and are widely considered to be the best treatment solutions, especially if water reuse is a requirement.

A key motivation for MBR technology implementation in the sector has been water recovery and reuse, motivated primarily by cost savings but also by water scarcity (and regulatory requirements relating to water conservation), security supply and strategic corporate planning. The key issues surrounding the water recycling in the sector concern the maintainance of hygiene and food safety standards using effective control measures. Since effluent reuse plants based on the two-stage MBR-RO within the sector have been established since the early noughties, the initial reticence associated with water recycling has largely been replaced with widespread acceptance. A final disinfection stage (chemical and/or UV) is normally sufficient to meet the required strict quality assurances for treated water reuse in the F&B sector.

3.2.1 Fermentation industries

Fermentation industries include malt houses, breweries, wineries and distilleries. Many examples of the different MBR configurations exist at full-scale, and implementation continues to grow in this sector.

3.2.1.1 Malting

Barley malt is a raw material used for brewing beer, for distillation of spirits such as whisky, and for producing malt extract and malt vinegar. In the malting process, barley is steeped in water, some of which it imbibes, and so initiating germination. The water may be changed two or three times during steeping and each successive change of steep water when discharged will contain decreasing amounts of 'pollutants' washed from the grain. Concentrations of BOD and TSS in the discharged steep water mainly depend on the volumes of water used. Generally, maltings produce relatively high volumes of low strength and readily-biotreatable wastewater, though the composition varies between sites (Table 3-1). There are now many MBRs installed in the maltings sector, mostly recovering water for reuse (Table 3-2), based on both pumped sMT MBR technologies (such as the Aquabio *AMBR*™ system, Section 4.3.3) and FS and HF iMBR technologies.

Table 3-1 Examples of maltings effluent quality

Parameter, mg/L	*Holland Malt, Eemshaven*	*Boortmalt, Sobelgra*	*Rahr Malting, Minnesota*
COD	1,500	1,880-2,100	-
BOD	1,350	700-930	620
TSS	<200	330-460	90
TKN, *TN*	50	*35-50*	47
$N\text{-}NH_4$	-	-	36
TP	10	13-15	-

Table 3-2 Examples of MBR maltings installations

Plant, location	*Flow, m^3/d*	*Config., Suppl.*	*Notes*
Simpsons, Berwick on Tweed, UK	1,400	sMT, Aquabio	Downstream RO, 630 m^3/d (45%) recovered
Muntons, Stowmarket, UK	40	sMT, Aquabio	Downstream RO, 30 m^3/d (75%) recovered
Bairds, Witham, UK	658	sMT, Aquabio	Downstream RO, 490 m^3/d (74%) recovered
Holland Malt, Eemshaven, Netherlands	1,400	sMT, Pentair	Downstream RO, 1,100 m^3/d (80%) recov.
Boortmalt, Sobelgra, Belgium	2,000	iHF, Koch	-
Canada Malting, Calgary, Canada	4,500	iHF, GE	Effluent reuse, cooling at local power station
Rahr Malting, Minnesota, US	8,500	iHF, Evoqua	Effluent discharged to local river

3.2.1.2 Brewing

Brewing operations comprise mashing (or lautering), boiling, fermenting, ageing and packaging. The quality and quantity of brewery effluent generated can fluctuate significantly due to the size and location of the brewery and the efficiency of the processing methods. The wastewater from brewing mainly consists of cooling and wash waters from process vessels, tanks, pipelines, filters and floors and from the bottle and cask washing and packaging lines. Volumes of effluent vary between breweries but are normally 4-10 times the volume of beer produced, typically containing 2,000-7,000 mg/L COD (Table 3-3), and this effluent is relatively easy to treat biologically.

Table 3-3 Brewery wastewater composition

Parameter	*General*	*Example installation (VA, US)*
COD, mg/L	2,000-7,000	6,900
BOD, mg/L	1,200-4,200	4,100
TSS, mg/L	200-1,000	600
TN	25-80	57
TP	10-50	15
T,°C	18-40	-
pH	4.5-12	-

For high organic loads anaerobic treatment is normally considered by the larger breweries, especially where space is limited. However, the anaerobic plants, which are mostly high-rate granular sludge technology, are sensitive to variable loadings and operating conditions and thus require specialist operational management to avoid possible plant failure. An example is a 9,500 m^3/d capacity plant in Virginia, US (Table 3.3), comprising staged anaerobic, pre-anoxic, aerobic and post-anoxic tanks plus 4 trains of 9 iFS cassettes (Kubota) to provide final effluent with BOD, TSS, N-NH_3, TN and TP concentrations of less than 5, 2, 1, 4 and 0.30 mg/L respectively. A smaller plant (2,300 m^3/d) in Venezuela employs a UASB upstream of an AL-sMBR, the UASB producing effluent COD, BOD and TSS concentrations of 306, 120 and 293 mg/L respectively. The 60-tube skid is claimed to operate at ~0.25 kW/m^3, comparable to the iHF configuration.

Aerobic treatment systems alone are widely preferred by many small-to-medium size breweries for their relative simplicity and more effective treatment with respect to nutrient removal, especially if discharge is to a sensitive watercourse or the treated water is to be reused. As with other industrial and municipal sectors, many examples exist of upgrades to existing SBR (sequencing bioreactor) plants to increase capacity without incurring an increased footprint. Examples include the Rothaus Brewery in Germany, whose treatment plant capacity was trebled by the retrofitting of a MICRODYN-NADIR *BIO-CEL®* based MBR technology (operating at a flux of 8-20 LMH), albeit with significant uprating of the blowers.

Reuse in brewing is reflected in the "water-to-beer ratio", which has been reduced to as low as 3:1 by the installation of MBR-RO systems in some cases (such as the 750 m^3/d plant at Shepherd Neame, the UK's oldest brewery, based in Faversham in Kent). A major challenge at many sites is the constraint on available space. This often leads to the selection of the sidestream configuration, since the skid can be adjusted to reduce the area footprint (to as little as a 1.5 m width in some cases) and the bioreactor can be operated at relatively high MLSS concentrations of up to 25 g/L.

3.2.1.3 Distilleries

There are many types of distilleries producing a wide variety of products by taking agricultural feedstock (such as grains, agave, molasses, etc.) and conducting processes such as cooking, fermenting, distilling, storage and maturation to blending and bottling the alcoholic beverages. Malt whisky, for example, is produced in pot stills from fermented washes made from 100% barley malt; grain whisky and spirits for gin and vodka are produced in continuous stills from fermented washes made from malted barley and other cereals, usually maize, and some forms of industrial alcohol, particularly for use in pharmacy and perfumery, are produced in stills from fermented molasses.

The composition of wastewater can vary depending on the type of distillery and its operations. In each type of distillery, the alcohol removed by distillation represents only a fraction of the volume of wash, and although a considerable quantity of the carbonaceous content is removed as CO_2 during fermentation, there remain considerable quantities of potentially polluting wastes for disposal. This is known as "pot ale" in malt distilleries and "spent wash" in others. Other wastes include "spent lees" which is liquor left over from the second distillation of spirit in malt distilleries, plus the CIP wastewater from washings of pipes, vessels and floors and sometimes condensate (if the site operates an evaporative dryer to concentrate the pot ale into a syrup for animal feed).

Anaerobic MBR technology has been successfully demonstrated for treating distillery wastes, including a number of immersed FS (Kubota) systems in Japan for treating the stillage from shochu distilleries, at loadings of 0.2-60 te/d, and sidestream (including Memthane) for treating stillage effluents generally. As with other beverage sectors, aerobic MBRs, often based on external pumped membranes, are generally preferred over anaerobic systems for relatively small flows of lower-strength effluents such as spent lees, washings and condensate. MBRs provide the added benefit of enhanced removal of copper (arising from copper stills) for meeting stringent discharge water quality limits.

There are two pumped sMBR systems installed (by WEHRLE) in Scotland for malt whisky distillery wastewater include those at the Glenallachie and Glenfarclas distilleries. These have both been operating for a number of years challenged with feedwater of ~3,000 mg/L COD and 2,000 mg/L BOD and generate treated water of <50 mg/L COD, <5 mg/L of BOD and <0.35 mg/L total copper. The operating flux is~100 LMH, and the sludge loading rate maintained relatively high (>0.2 kgCOD/kgMLSS, 10,000-20,000 mg/L MLSS) so as to minimise copper accumulation in the MLSS.

3.2.1.4 Wineries

Winery processes normally involve the crushing and pressing of grapes, fermentation, ageing followed by blending and bottling. Wineries normally require some form of wastewater treatment, with roughly 10 m^3 of wastewater per m^3 of wine produced. Vintage wine production operates for only three months of the year, during which time the bulk of the wastewater (60-70%) is generated. The rest of the time wastewater is significantly lower in strength and volume.

Since many wineries are located in rural locations the treatment systems typically involve a series of lagoons followed by irrigation to land. SBRs are widely used providing a relatively simple and low cost solution for aerobic treatment. However, with increased site capacity and stringent legislation for direct discharge to the environment, accompanied by growing concerns over future water supply, there is increased interest in advanced biological treatment such as MBR systems that offer the potential for reuse of the treated water.

Many small (20-200 m^3/d) iMBR installations for wineries exist, particularly in Spain and Italy. These typically operate at a flux of ~15 LMH and generate final treated water BOD and COD of <25 and 50-300 mg/L respectively according to seasonal changes in the feedwater flow and quality (Table 3-4). MBRs based on the Kubota system are also installed at some wineries in California, where typical influent water quality values during the crushing season are 500-12,000 mg/L BOD, and 40-800 mg/L TSS, the MBR permeate water being generally between 2 and 10 mg/L BOD.

Table 3-4 Italian winery, 2007 water quality data

Parameter	*Unit*	*Jan*	*Aug*
Flow	m^3/h	15	44
TSS	mg/L	536	780
COD	mg/L	1,327	3,978
BOD_5	mg/L	610	2,295
TN	mg/L	36	32
TP	mg/L	12	22
pH	-	7.7	10
Temperature	°C	12	22

3.2.2 Dairy industries

Dairy products include cream, cheese, butter, yoghurt and ice cream, as well as various condensed and powdered milk and whey products all produced from milk. The wastewater characteristics vary depending on dairy operations and the overall site efficiencies, but contain mostly organic compounds from lost product with ~40% of the total COD due to fats, ~30% due to proteins and ~30% due to lactose. COD concentrations generally lie between 1,000 and 5,000 mg/L for milk and cheese processing wastewaters.

The use of MBR technology for treating dairy wastewater has been successfully demonstrated for many years, with reference plants for all the main technology configurations and from small- to large-scale. These are mostly based on aerobic treatment, although many plants include anoxic and sometimes anaerobic stages for N and P removal. MBR technology application can be particularly challenging in some dairy sectors due to the variable feedwater quality coupled with its relatively high fouling propensity attributed mainly to its FOG and protein content and/or the likely precipitation of scale due to high calcium levels. Early dairy effluent treatment MBR plants were prone to excessive fouling or scaling, requiring more frequent cleaning or pre-treatment. As such, the pumped sMBR technologies offer the advantage of more precise hydrodynamic control and a simplified, robust CIP protocol made possible by the isolation of the membrane from the sludge during chemical cleaning.

In general, regardless of the technology used or the feed COD concentrations, COD is generally removed to well below 100 mg/L in the permeate, often below 50 mg/L, with BOD and ammonia levels usually below 5 mg/L. This reflects the biodegradability of these effluents, with BOD/COD ratios usually >0.6. For sMBR systems fluxes in excess of 150 LMH can be sustained provided chemical cleaning is adequate to control scaling.

3.2.3 Carbonated soft drinks and juice beverages

The making of soft drinks typically involves the production of concentrate, syrups and flavours/infusions which are then blended with treated water and packaged into the various container types. Water is used for the cleaning of various vessels and lines. Most facilities generate many types of product and water is used during the rinse cycles to switch between products, which is a major contribution to the total volume and loading of the wastewater from the larger bottling plants. There may also be product losses from the syrup production and bottling lines which include washing, rinsing and, for some producing juice drinks, pasteurisation.

The main pollutant from bottling or beverages plant is BOD and COD, which is mostly all due to dissolved organics. Typically, the mixed wastewater will have a BOD of 2,000-3,000 mg/L and COD of 3,000-5,000 mg/L with a relatively low TSS of 100-400 mg/L and TDS values mostly <3,000 mg/L. However, there are MBR applications with effluents of high syrup or juice content where the influent wastewater BOD may be as high as 20,000 mg/L. One such effluent, at a bottling plant in North America, is treated by a pumped sMBR which achieves an effluent BOD of <500 mg/L (>97% BOD removal) and operates well despite the feed containing 125-150 mg/L FOG at times. For such high loads anaerobic pre-treatment is often viable.

An example of the use of anaerobic pre-treatment is at a juice and sport drinks processor in Florida US, which had its existing anaerobic treatment (based on the ADI-BVF® reactor, a proprietary low-rate anaerobic system) upgraded in 2007 to include an aerobic iFS MBR (Kubota) with downstream RO polishing to meet more stringent discharge standards on final BOD and TDS. In this case the feedwater contains 3,000 mg/L COD, 1,700 mg/L BOD and 2,200 mg/L TSS, the COD being reduced to 450 mg/L by the ADI-BVF® reactor and then to 35 mg/L by the MBR. In this case the MBR is required to treat only half of the effluent to attain the stipulated discharge water quality standard.

3.2.4 Cereals and snack foods

Companies in this sector produce ready-to-serve cereal, breakfast snack bars and cereals such as oatmeal that must be cooked prior to eating. The most popular cereals are made with corn, wheat, oats, mixed grains, or rice mostly as flaked, puffed, shredded or shaped products. Many cereals are sweetened by adding malt, white sugar, brown sugar, corn syrup or concentrated fruit juice. They may also include flavours such as chocolate and cinnamon.

Cereal facilities can produce significant amounts of wastewater mostly from the cleaning of equipment (i.e. cookers, conveyors, rollers, extruders, flavour tanks, pipework, etc.), normally sterilised with pressurised steam, plus water discharged from scrubbers and site utilities. The influent wastewater quality is typically 300-500 mg/L TSS and 2,000-3,000 mg/L COD, but it may be much higher (possibly 2-3 times more) from some facilities if there are more product losses, use of sweetener coatings and frequent product changes entailing more cleaning operations. As such, there may be a requirement for primary treatment, normally using DAF-based clarification, to reduce the higher influent TSS of 1,000-2,000 mg/L and associated COD of 5,000-10,000 mg/L.

Early MBR plants installed for treating cereals manufacturing wastewaters include the sMBR at the Kellogg plant in Manchester, UK, in 2004. The original installation was designed for 1,500 m^3/d (at 4,000 mg/L COD) but was later upgraded to a 2,100 m^3/d plant treating 7,100 mg/L of COD. The upgrade included the addition of a fifth bank of membranes and the conversion of the balance tank into a bioreactor using pure oxygen for enhanced aeration to address the additional load. The upgraded MBR provides 99% removal of the influent COD, with permeate COD and BOD values of 10-100 and <10 mg/L respectively with downstream RO for effluent reuse. A similar pumped sMBR technology has been installed at the General Mills facility at Covington, GA, which achieves a comparable effluent quality from a similar feedwater. In this case, 50% of the effluent is recycled for use in non-potable factory applications (such as the dust scrubbers) and as demineralised feedwater for the boilers, significantly reducing the total site operating costs associated with chemicals use and blow-down wastage.

An unusual example of a small (95 m^3/d) AnMBR treatment followed by an AeMBR exists at another Kellogg facility. In Pikelville, Kentucky in the US the existing anaerobic and aerobic wastewater treatment facility was upgraded to a two-stage AnMBR-AeMBR process to treat water having a mean COD of 34,500 mg/L, including a 19 m^3/d sanitary wastewater flow sent directly to the aerobic system. The AnMBR provides a treated effluent 160 mg/L COD on average which feeds the AeMBR. The permeate then undergoes a series of chlorination and dechlorination steps, the complete system maintaining >99.5% COD removal and an average effluent COD concentration of 62 mg/L. Other parameters (NH_3-N, FOG, E. coli, etc.) all meet the strict discharge limit, allowing the final effluent to be discharged directly to the local creek and so avoiding costs of discharge to the municipal sewers.

3.2.5 Potato and starch industry

Wastewater and organic wastes generated in potato processing (for producing potato chips, French fries, dehydrated mashed potatoes, potato flake, potato starch, canned potatoes, pre-peeled and diced potatoes and related foods) result from various washing stages, peeling, and other processing operations. The waste is first treated to recover valuable by-products, such as starch, oils and grease, then settled prior to biological treatment which now includes anaerobic and aerobic MBR technologies which provide the opportunity for water reuse.

The wastewater characteristics depend on the processing operations and end product, but the wastewater is normally high in concentrations of organic materials like starch and proteins and so very prone to fermentation and frothing. It contains relatively high concentrations of organic and inorganic nitrogenous compounds and phosphates. Processing based on heat treatment steps such as blanching, cooking, caustic treatment and steam peeling tends to produce gelatinised starch and coagulated proteins, incurring a much higher COD load. Other constituents include dirt, caustic, oils and fats, cleaning and preserving chemicals.

An average-sized potato processing plant producing French fries and dehydrated potatoes can create a waste load equivalent to that of a city of 200,000 people, with around 17 L of wastewater generated per kilogram potatoes. Wastewaters generally contain 4,000-18,000 mg/L COD with a BOD/COD ratio generally above 0.5 and TKN and N-NH_4 levels of 150-500 and 50-250 mg/L respectively. TSS levels are comparable to those for COD, with these solids including easily settleable sand and soil particles from the initial potato washing stage. As such, screening, equalisation and primary sedimentation can reduce the concentrations to 2,500-5,000, 1,500 -3,000 and 250-500 mg/L COD, BOD and TSS respectively.

Potato processing wastewaters are slightly more challenging for MBR treatment than other food wastewaters because of their high colloid and gelatinised starch content, making them more highly fouling in nature. Also, the presence of abrasive inorganic particles can be onerous to some membranes. MBRs have nonetheless been used for potato starch production since the mid-1990s (for example the iHF MBR plant for ABR Wietzendorf, Germany, which was originally a 480 m^3/d plant but has since quadrupled in capacity). There are now a number of MBR installations at various types of potato processing facilities including the iHF MBR installation at Farm Frites in Belgium (potato starch), a ~200 m^3/h iHF plant for Basic American Foods (Section 5.1.7) and a ~100 m^3/h iHF MBR plant for Frito Lay in Casa Grande, Arizona which produces potato chips and snacks. An external tubular MBR system was installed by WEHRLE at KP Foods Ltd. in the UK (2009) to process pre-treated wastewater and provide 65 m^3/h of water for reuse in the initial potato washing stages.

3.2.6 Salads and vegetable products

The market for prepared salads and vegetables has increased substantially in recent years, and there is also the large established market of canned vegetables such as beans and soups. Both these activities produce large volumes of wastewater from the various washing stages and in the fluming (the transport of produce using water) between the various operations. Also, there are some facilities that discharge higher strength waste liquors associated with heating processes.

Modern vegetable processing facilities often already include internal water recycling and inter-stage water reuse opportunities (for example simply by using counter-current inter-stage washing). Older facilities are often less efficient in terms of water usage and waste generated. The quality of the wastewater discharged depends on the produce and the various operations, including any internal recycling of water, and the presence of high-strength wastewater associated with heating operations such as cooking and blanching, for example in the production of soups and tinned vegetables.

For a prepared salads facility, a typical wastewater quality discharged (following screening) has a COD, BOD and TSS of 1,500, 800-1,000 and 100-500 mg/L respectively. In contrast, vegetable blanching liquors can have CODs an order of magnitude higher than this (similar to potato processing wastewaters) with the combined wastewater having a COD of 2,500-5,000 mg/L for a cannery or soup production facility.

There are now very many MBR-RO installations associated with this industrial sector with over 10 years of reliable treatment and recycling of water meeting the required stringent hygiene and potable water quality standards. One of the early applications of the two-stage MBR-RO process for treating wastewater from a prepared salads producer is at Kanes Foods (Section 5.1.13), which was originally built as a classical pumped sMBR in 2001 and supplemented with a low-energy *AMBR LE™* process (Aquabio/Freudenberg) in 2010. Both separate treatment schemes encompass screening, equalisation and DAF as pre-treatment, and the MBR permeate has an average COD below 20 mg/L.

3.2.7 Confectionery

The wastewater properties in confectionery factories vary widely depending on the range of products and seasonal impacts. Generally, the wastewater is of relatively high strength, mostly due to dissolved organics (BOD and COD), but low in TSS and nutrients which may therefore demand the addition of N and P for optimised biological treatment. There may also be higher FOG levels if the facility makes confectioneries using chocolate. A popular confectionery product type is jellied fruits, sweets and gums. These are made using sugar, syrups, flavours, pigments and gums that are all blended and the mixture filled into cups or various shapes moulded on a conveyor belt of starch powder, and then often coated.

The main sources of wastewater are from the washing of various pieces of equipment and product lines, and spilling of sugar, syrup or overflow of blended liquor and coatings from sealing machines. For relatively low volumes, wastewater can be highly variable with high COD concentrations. Viscous sludge bulking generally tends to occur in treating confectionery wastewater (due to polysaccharides), making sludge separation difficult when using conventional treatment. In such cases MBR technology offers distinct advantages over alternatives such as the conventional activated sludge (CAS) process. Operation of the MBR under optimised conditions by adjusting the sludge age and MLSS concentration to maintain an appropriate BOD-MLSS load (0.2-0.4 kg BOD per kg MLSS per day) can retard the growth of bulking filamentous micro-organisms.

There are now a number of MBR systems installed by Cadbury at various confectionery plants in Europe (Table 3-5), mostly based on sMBR technology but with some using the iMBR configuration. One of the early sMBR projects (2005) was for the Cadbury site at EuroCandy in Lille, France, based on Pentair *X-Flow* UF modules and installed by the French engineering company Proserpol. It was, however, challenged with many operational issues in the early stages of its operation, in particular the frequent membrane tube blockage and control of the variable influent feed. These were linked to the accumulation of inert solids (starch powder from production) and the return of excess polymer from the centrifuge sludge dewatering process that significantly increased the viscosity of the mixed liquor in the UF loop. There were also some operational issues reported with the submerged MBR system in Beirut, adversely affecting both membrane flux and aeration capacity, mainly related to plant operating conditions and control of the highly variable influent feed quality and lack of essential nutrients. Other Cadbury sMBR plants (supplied by WEHRLE) are reported to have been working effectively over many years with typical influent COD values of 8,000-10,000 mg/L and COD removal of 90-99% to meet final discharge requirements.

Table 3-5 Cadbury MBR plants

Production Site	*Capacity m^3/d*	*COD load*	*Bioreactor volume, m^3*	*Membrane area, m^2*	*Design/supply by:*
Beirut, Lebanon	100	200	200	600	Grossimex/Zenon
EuroCandy, France	120	1,200	350	81	Proserpol
Bucharest, Romania	50	250		30	WEHRLE Umwelt
Valladolid, Spain	400	4,000	900	224	WEHRLE Umwelt
Novgorod, Russia	200	1,600	2 x 175	108	WEHRLE Umwelt

3.2.8 Edible oils and spreads

The edible oils industry produces cooking and specialty oils and fats, margarines, mayonnaise, shortenings and whipped toppings for sale to other food processors, food service and retail markets. Vegetable oils are produced from rapeseed, corn and soybean, by extraction and then refining of the oils. In the refining process, dust, saccharides, proteins, gummy substances, fatty acids, pigments, smelling substances and other such items are removed. Phosphoric acid, sodium hydroxide and water are used in the process, and filter aids dosed to improve filtration and refining. Water consumption volume varies with the production capacity and type of vegetable oil.

Vegetable oil processing wastewater generated during oil washing and neutralisation tends to have a high FOG content and COD levels in the range 5,000-10,000 mg/L. The wastewater may also be high in TSS, organic nitrogen and sometimes contain pesticide residues from the treatment of the raw materials. Treatment accordingly includes grease traps, skimmers or oil-water separators for floatable solids; flow and load equalisation; coagulation and DAF clarification of the colloidal and suspended solids. Generally, 60-80% of the total organic load can be removed during the primary DAF stage. Some sites have comprehensive wastewater treatment plants that include biological treatment, typically with anaerobic followed by aerobic treatment, along with biological nutrient removal of the total nitrogen and phosphorus prior to final discharge. MBR technology is being more widely considered for both anaerobic and aerobic treatment of these wastewaters.

An early application of membrane technology was for a large facility producing oilseed products (ADM Decatur, Illinois, US). This project used GE *ZeeWeed* membranes for tertiary treatment of 18,900 m^3/d of wastewater (upgraded in 2006 to 22,700 m^3/d), for reuse in the cooling towers. Another GE reference site is for an oils and shortening facility in Cincinnati, US, treating 3,300 m^3/d of wastewater, with plans to upgrade to 5,500 m^3/d, with treated water to be reused for cooling water make-up. The influent to the MBR has typical COD, BOD, TSS, FOG and TP values of 800-2,000, 600-1,000, 70-200, 50-150 and 30-50 mg/L respectively. The MBR permeate contains <10 and <0.5 mg/L BOD and TP respectively, with P biological and supplementary chemical removal. A recent project by Kubota, based on their AnMBR process, for a palm oil processing plant in Malaysia was commissioned in 2013. Other MBR projects are being considered by the palm oil industry.

3.2.9 Meat and poultry processing

The meat and poultry processing industry covers a wide variety of processing operations from small to large plants of first processors (slaughterhouses) to further processors making finished products (e.g. bacon, sausages, cured meats, and 'ready-to-cook' meats such as poultry). This segment of the food processing industry offers more difficult waste streams to treat, with mostly high BOD, FOG, and blood by-products from the killing, washing and rendering processes. It is by far the most regulated and monitored sector due to the possibility of disease spread by pathogenic organisms. There are now strict water quality standards for the meat and poultry products (MPP) industry, in particular with regard to the discharge limits for nutrients such as ammonia, total nitrogen (TN) and phosphorus and for the required quality of water for reuse.

Most modern MPP processing plants, as in many other food processing activities, are high water users and there is increasing pressure on reducing the total water consumption by exploring reuse. A typical meat packing plant would require ~13-18 m^3 of water per ton of product and as much as 20-40 L of water to process one average-sized 2.5 kg chicken. As such, it is not unusual for a typical MPP processor to generate 1,000-5,000 m^3/d of wastewater.

The application of MBR technology has been proven for many years in the MPP industry with both large- and small-scale plants based on the various membrane configurations and including RO for water reuse. One early application of the sMBR process was by WEHRLE for a rendering operation (now Argent Energy) in Motherwell, Scotland. This plant was commissioned in 2002 to treat 450 m^3/d of wastewater with influent COD, BOD and ammonia values of 5,200, 3,650 and 700 mg/L, respectively. The treated effluent has COD, BOD and ammonia values of <100, <5 and <50 mg/L, respectively and complies with the required sewer discharge limits.

The plant at Cooperl, Lamballe in France includes an iHF MBR process with downstream RO for water reuse from pork processing, with Phase I (2003) treating 1,040 m^3/d, increasing to 1,360 m^3/d in Phase II (in 2004). This site now employs Koch Membrane Systems technology (*PSH 500*, Section 4.2.7) installed in 2007 to process 960 m^3/d of wastewater, with 8 modules providing a total membrane area of 4,000 m^2 (and hence a net flux of 10 LMH). The effluent load to the MBR and the treated effluent quality are shown below. The COD, BOD, TN and TP are respectively reduced from 1,780, 929, 288 and 5.7 to <50, <10, <50 and <1 mg/L in the MBR effluent. The RO permeate contains <10 mg/L COD, <1 mg/L TN and <1 CFU/100 mL.

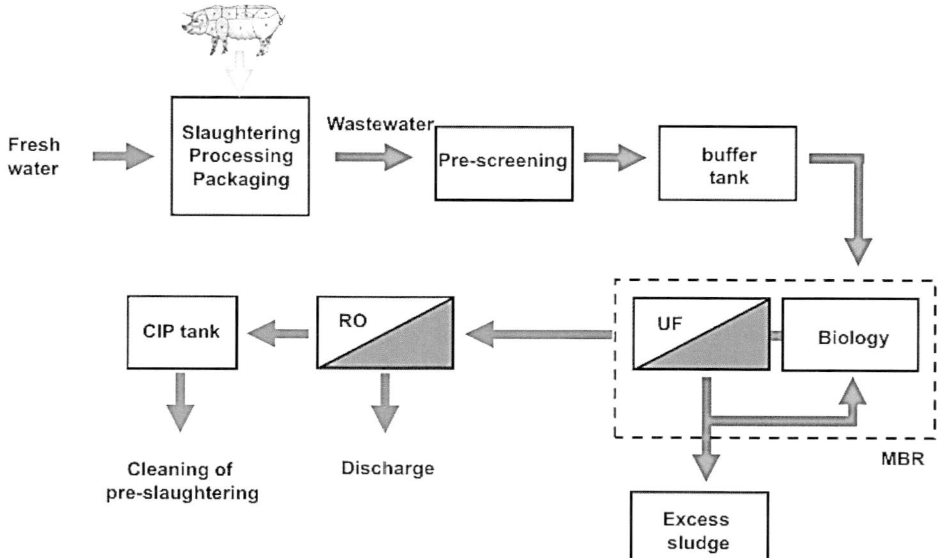

Figure 3-1 The slaughterhouse effluent treatment scheme at Cooperl

A large meat processor in Bavaria, Germany has installed a full-scale MBR plant based on the *Vacuum Rotation Membrane* (*VRM®*) bioreactor system from Huber Technology (Section 4.1.6), commissioned in 2008 after extended pilot trials to prove the process. The MBR plant treats on average 1,200 m^3/d of wastewater, with a maximum flow of 1,600 m^3/d. It includes coarse screen, fine screen, equalisation, and DAF primary treatment that removes ~50% of the influent COD loading. The MBR is designed to treat 1,920 kgCOD/d with 3 lines, each equipped with one 30/400 *VRM®* unit, providing a total area of 2,400 m^2 and thus a mean flux of 13 LMH. The MBR influent COD is 1,750 mg/L on average, with 5 mg/L of ammonia and 20 mg/L of phosphate. The mean permeate COD is 20 mg/L and consistently meets the required water discharge standard to the local creek – and even water reuse quality standards.

3.3 Petroleum industry effluent

The petroleum industry comprises three elements: exploration, refining and petrochemical, where refining and petrochemical production may be combined. Exploration concerns the abstraction of mineral oil and gases from underground reservoirs. Refining mainly concerns the separation of the crude oil into useful fractions and petrochemical operations are those involving chemical modification of these fractions into further products.

3.3.1 Exploration wastewater

The single largest volume wastewater generated in the petroleum industry is from oil exploration. This effluent, "produced water" (PW), results from the use of environmental waters – most often seawater – to displace the oil in the reservoir. PW contains a number of different chemical species which are onerous with respect to either their environmental impact or to abstraction operations (due, for example, to scale or hydrate formation). These species, which can vary significantly in concentration both regionally and temporally, can be categorised according to their origin and/or chemistry (Fig. 3.2a). The extent to which they must be removed depends on the management strategy, and the treatment technologies selected for this depend on whether the installation is based onshore or offshore (Fig. 3.2b). For onshore installations, where footprint is not a critical factor, the relatively low-energy, simpler, high-footprint technologies can be employed. This may include the same biological treatment technologies as routinely applied to other industrial effluents, including refinery and petrochemical effluents.

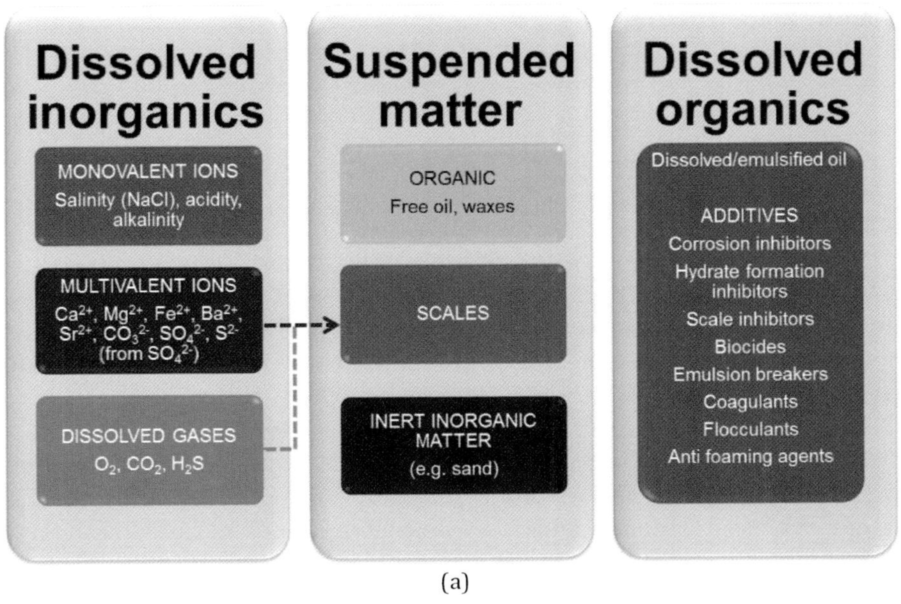

(a)

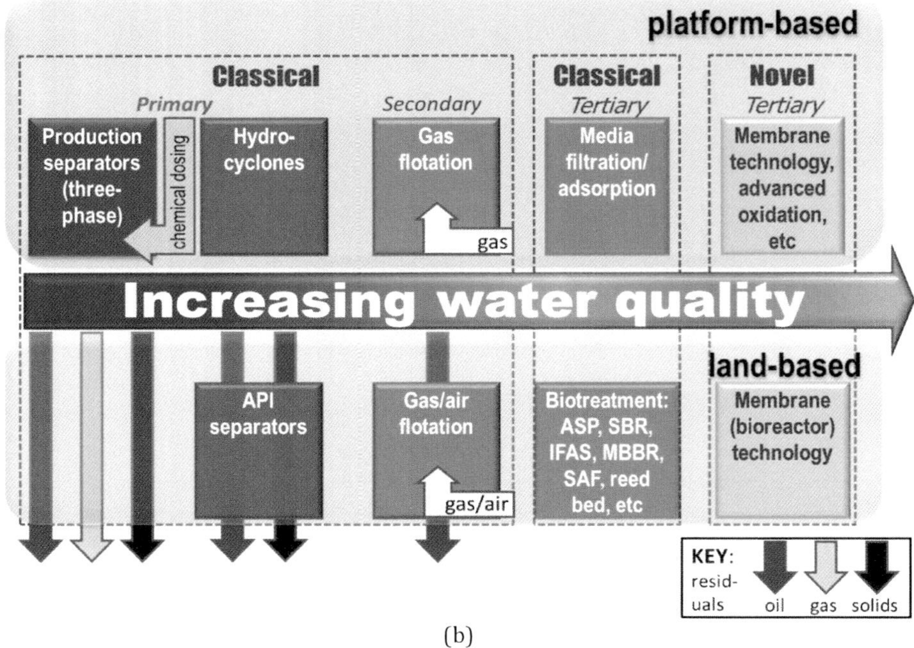

(b)

Figure 3-2 (a) Produced water (PW) primary constituents, and (b) Treatment options offshore (for PW) and onshore (PW and refinery wastewater) indicating potential MBR option.

The ranges of concentration of the key constituents (Table 3-6) vary widely, and generally the exact composition with reference to the additives is not known and/or considered proprietary by the industry. Thus, whilst the biodegradability of the mineral oil components can be quantified, assessment of biodegradation of the additives can be challenging. The application of MBR technology to actual PW wastewaters (Table 3-7) appears to be still at the development stage (Kose et al., 2012; Pendashteh et al., 2012, Sharghi et al., 2013), although full-scale sMBR installations have apparently been implemented (Section 5.2.5).

3.3.2 Refinery wastewaters

3.3.2.1 Refinery wastewater origins

Refineries generate products from crude oil (or "crude") by thermal fractionation: separation of the crude constituents takes place by virtue of their differing boiling points. Wastewater is generated from the refining process from a number of sources including (Table 3-8):

- tank bottom draws
- desalter effluent
- stripped sour water, and
- spent caustic.

Entrained water in the crude derives from the oil well extraction process and/or from ingress during transportation. It is typically removed as storage tank bottom sediment and water (BS&W) or in the desalter, which is a key component of the crude oil processing at the refinery, and forms part of the wastewater. A significant effluent stream derives from where pre-softened or stripped sour water has been in contact with hydrocarbons. Wastewaters generated from operations from where no direct contact with hydrocarbons arises include residual water rejected from boiler feedwater pre-treatment processes, water produced from (i) regeneration of ion exchange resins in zeolite softeners and demineralisers, and (ii) blowdown (the concentrate stream) from cooling towers and boilers. There is also likely to be minor contamination of stormwaters from run off, as well as minor flows from laboratory discharges, washing and sewage.

Table 3-6 PW quality from oil fields and gas fields, all in mg/L other than pH (adapted from Fakhru'l-Razi et al., 2009)

		COD	*BOD*	*TSS*	*TDS*	*N-NH$_4$*	*pH*	*Cl*	*Ca*	*HCO$_3^-$*	*O&G*[a]	*Phenol*
Oil	Min	-	-	1	-	10	4.3	80	13	77	2	0
	Max	1,220	-	1,000	-	300	10	200,000	26,000	4,000	565	23
Gas	Min	2,600	75	8	2,600	-	3.1	1,400	9,400	-	2	-
	Max	120,000	2,900	5,500	360,000	-	7	190,000	51,000	-	60	-

[a]O&G oil and grease

Table 3-7 Summary of MBR performance for treatment of petroleum wastewaters (adapted from Lin et al., 2012)

Feed	*Membrane*	*Flux* LMH	*Bioproc.* config	*$V_{reactor}$* L	*COD_{in}* kg/m^3	*OLR* kgCOD/m^3.d	*HRT* h	*SRT* d	*F:M* kgCOD/kgVS S.d	*MLSS* g/L	*T* °C	*% rem* COD	*Reference*
PW	iHF, 0.1µm	10	Ae	5.1	1.5-3	13-26	2.7[b]	30-"inf"	0.25-0.45	2-16	20	67-83[a]	Kose et al., 2012
PW analogue/ real	PVDF sMT, 200kDa	75-95[e]	Ae SBR	-	0.56-6.8	0.28-3.4	24-96	"inf"	0.15-0.57	1.6-7.9	30	97-99[c]	Pendashteh et al., 2012
Petrochemical	PVDF iHF, 0.04µm	10-18	Ax/Ae	2,000/ 2,200	0.074-0.223	0.08-0.50	10.7-18.3	50-90	0.042-0.11	3.0-4.8		69-87	Di Fabio et al., 2013
Petrochemical	iFS chlorin. PE, 0.4µm	10-12.5	Ax/Ae	18/30	0.720-1.59	0.9-3.0	13-16	25	0.12-0.23	8.6-9.6	26	85-95[d]	Qin et al., 2007
Refinery	Ceramic sMT, 0.2µm	50-120[e]	Ae	20	0.37-2.3	0.024-0.067	17-34	-	-	3-5.5		~94	Rahman & Al-Malack, 2006
Refinery	iHF	15-17.5	Ae	4.4	0.4-1.05	0.74-1.72	10.0	"inf"	0.27-0.77	2.1-10.4	~25	41-67	Viero et al., 2008
Refinery	iFS PVDF 0.08µm	-	An/Ax/Ae	-	0.072-0.296	-	-	-	-	-	13-17	89-98	Zhidong et al., 2009
Oil-water analogue	sMT PVDF 15kDa	40-100[e]	Ae	11	0.5-3	0.82-9.82	6.7-13.3	-	0.26±0.54 g	2.5-30	35	93-98[f]	Scholz and Fuchs, 2000
PW	iFS chlorin. PE, 0.4µm	-	Ae	0.75	0.6-1.8	0.3-0.9	48	80	-	1.1-5.2	30	75-95	Sharghi et al., 2013

a Generally 80-85% independent of COD_{in}
b Calculated from other reported data in publication
c Decreases to 90% on increasing salinity from 35 to 250 g/L
d Insignificant impact of HRT between 13 and 19 h on COD_{out} (40-65 mg/L)
e Highly dependent on membrane fouling condition, TMP and CFV
f Other than at lowest OLR of 0.82 when removal was 77%

OLR Organic loading rate
HRT Hydraulic retention time
SRT Solids residence time
F:M Food/micro-organism ratio
MLSS Mixed liquor suspended solids concentration
iHF Immersed hollow fibre
iFS Immersed flat sheet
sMT Sidestream multi-tube
chlorin PE Chlorinated polyethylene
PVDF Polyvinylidene difluoride
Ae Aerobic
Ax Anoxic
An Anaerobic
"inf" "infinite" (no sludge wasting: SRT determined by sludge sampling)

Table 3-8 Refinery effluent stream water quality, mg/L (IPIECA, 2010)

Parameter	*BS&W*[a]	*Desalter*	*Stripped sour water*	*Cooling tower blowdown*
COD	400-1,000	400-1,000	600-1,200	150
Free HCs	Up to 1,000	Up to 1,000	<10	<5
SS	Up to 500	Up to 500	<10	Up to 200
Phenol	-	10 – 100	Up to 200	-
Benzene	-	5 – 15	negligible	-
Sulphides	Up to 100	Up to 100	-	-
Ammonia	-	Up to 100	-	-
TDS	High	High	Low	Intermediate

[a]Tank bottom basic sediment and water

The principal water stream in a refinery is the cooling water (CW), which makes up about 50-55% of all the water in a refinery. At times CW can by-pass the WwTP to reduce its hydraulic loading provided the CW quality is appropriate for discharge. If contamination from a leak is detected then CW is rerouted back to the WwTP. In addition, CW may be used for dilution of high-COD waters if they are otherwise by-passing the WwTP.

Tank bottom draws

When the crude is stored in large tanks, the settled BS&W requires periodic removal to prevent its accumulation which would otherwise reduce the storage capacity. The extracted water is discharged to the effluent treatment plant, either directly or following sedimentation of the solids in a dedicated tank where the residual oil is also skimmed. Level indicators are installed to ensure that only the aqueous phase is extracted. BS&W is high in COD and can contain free oil and grease if not correctly drawn off for discharge.

Desalter effluent

Desalting is typically the first operation in a refinery after the crude unit, which removes the basic sediment and water (BSW). The desalter is intended to both prevent plugging and fouling of downstream separation process equipment by salt deposition, and reduce corrosion caused by the formation of HCl from the chloride salts during processing. As well as dissolved impurities the desalter also separates out solids, such as sand, drilling mud and paraffin waxes, and removes sulphurous and phenolic species. As the crude oil is pumped to the desalter it is pre-heated and mixed with 3-9% wash water which dissolves the salts in the oil. A mix valve upstream of the desalter is used to create an emulsion of the two liquids, which then begins to separate out into wash water and cleaned crude oil in the desalter to an extent dependent on factors such as retention time, temperature, mixing efficiency and pH. De-emulsification is achieved using electric grid technology, comprising large electrically charged plates which cause the water molecules to coalesce into droplets which then sink to the tank bottom. The treated oil is pumped off the top of the vessel while the effluent water is pumped off the bottom to waste treatment. Often the refinery will have a second-stage desalter to remove any residual salts from the dehydrated crude oil.

Inorganic salts are present in crude as an emulsified solution of salt, predominantly sodium chloride, originating from the oil reservoir formation and/or injected seawater. The concentration of water in the crude varies widely but is normally in the range 0.1–2.0% by volume. The salts contained in the aqueous phase, mostly in the form of chlorides of sodium, magnesium and calcium, range from 0.03 to 0.7 kg salt per m^3 of crude. Drilling muds, which are materials used to aid drilling by providing cooling and lubrication, may also accumulate in the desalter; their intermittent removal ("mud wash") can cause significant shock loads of suspended solids and hydrocarbons in the desalter effluent stream.

Wastewater quality from desalting varies widely according to the pH (with higher pH levels resulting in greater emulsification), the separation of the water and crude and the frequency and effectiveness of the mud wash. Mud washes can be continuous or intermittent, with continuous being preferred to avoid shock loads of organic matter into the WwTP. Organic contaminants also derive from the BSW from the crude unit.

Sour water
Sour water is condensed steam that has been used to control applications such as catalytic cracking. It contains hydrocarbons, TKN and mineral constituents such as ammonia (NH_3, contributing to the TKN) and hydrogen sulphide (H_2S), which are then typically stripped using steam down to levels which may be as low as <30 mg/L NH_3 and <1 mg/L H_2S. The stripped sour water may then be suitable for reuse within the installation, the excess being sent to the effluent treatment plant. Refineries which include processes such as catalytic cracking generate more sour water than simpler refineries, and sour water from these sources contains phenols and cyanides. Segregation of these waters for stripping allows the stripped sour water to be reused as desalter wash water, allowing the residual phenol to be extracted (by up to 90%) into the crude and so reducing the load of phenol to the wastewater.

Spent caustic
Caustic soda solution is used to extract acidic constituents from hydrocarbon streams, generating a spent caustic solution containing H_2S, phenols, organic acids, hydrogen cyanide and carbon dioxide. This stream is either intermittently or continuously partially replenished with fresh caustic, with the surplus discharged to the effluent treatment plant via product storage tanks. This stream may also be used in the fluid catalytic cracking unit to scrub sulphurous oxide (SO_x) gases, but may then cause heavy foaming in the fluid catalytic cracking unit.

Crudes having a high naphthenic acid content can be challenging, since when this compound is extracted into the caustic solution it forms naphthenates which are especially refractory to biological treatment (i.e. biorefractory). In such cases advanced oxidation, such as wet air oxidation at high temperature and pressure, may be required. Otherwise, removal of the H_2S and phenols may require acidification to below a pH of 3 to allow stripping of the H_2S and separation of the phenol. H_2S otherwise poses a significant safety hazard.

3.3.2.2 Refinery wastewater treatment

Refinery wastewater quality varies significantly (Table 3-9) according to the process cycles. Its treatment is generally based on classical activated sludge treatment, usually with an initial flotation sequence to remove the oil. The simplest flotation device is the American Petroleum Institute (API) separator, the "workhorse" in any refinery for the separation of oil/water and solids, which allows both settlable solids and large oil droplets (>150 μm) to be removed by up to 90%. This primary step is then often followed by clarification. This may comprise corrugated plate separators preceded by coagulation/flocculation and followed by either dissolved air flotation (DAF) or induced gas/air flotation (IGF/IAF). These technologies target much smaller oil droplets - 10-25 μm - and reduce the suspended oil concentration to around 25-50 μm.

Flotation, along with the increasingly employed electrocoagulation process, is most effective (in terms of % removal) for high suspended oil concentrations, such as those arising in the desalter and BS&W effluents. Such effluents, along with the spent caustic, also have a considerably higher salt content than the remaining effluent streams. It is therefore desirable to treat these three streams separately from the remaining low-TDS streams to allow both pre-treatment for oil removal and segregated biological treatment of high-TDS effluent. Since segregation is rarely employed significant shock loads arise in refinery effluents from dissolved salt and oil, in particular from sub-optimal electrical coalescence (grid technology) or intermittent discharge of the mud wash from the desalter.

Whereas biological treatment of PW is still at the developmental stage, it is routinely employed for refinery and petrochemical effluents (Ishak et al., 2012) where the application of MBRs has also been explored (Rahman and Al-Malack, 2006; Qin et al., 2007; Viero et al., 2008; Zhidong et al., 2010; Di Fabio et al., 2013). Data reported from these studies (Table 3-9) have indicated organic contaminant removals, expressed as chemical oxygen demand (COD), generally in the range of 84-99%, with fully optimised systems achieving >95% COD removal as well as complete nitrification (Qin et al., 2007; Zhidong et al., 2009; Di Fabio et al., 2013). Reported results indicate COD removals to vary little with the hydraulic retention time (Scholz and Fuchs, 2000; Pendashteh et al., 2012), but strongly dependent on the feedwater composition and, in the case of nitrification, pH: a decrease in pH levels to below 5.8 has been shown to reduce nitrification to as low as 80% (Zhidong et al., 2009). Biotreatment may also be enhanced by the addition of powdered activated carbon to the bioreactor, which helps retain the dissolved organic matter and thus extend the treatment time.

Table 3-9 Refinery effluent composition examples, mg/L (adapted from Diya'uddeen et al., 2011)

pH[1]	*COD*	*BOD*	*O&G*	*SS*	*NH_3*	*Phenol*	*S^{2-}*	*Reference*
7.0	300-600	150-360	<50	<150	15	-	-	Ma et al., 2009
8.0	80-120	40	23	23	-	13	-	Abdelwahab et al., 2009
6.6	600	-	120	120	-	-	890	El Nass et al., 2009
8.4	220	-	-	-	-	-	22	Altas & Büyükgüngör, 2008
6.5-7.5	170-180	-	420-650	420-650	-	-	-	Saien & Nejati, 2007
-	300-800	150-350	100	100	-	20-200	-	Al Zarooni & Eishorbagy, 2006
6.7	200	-	-	-	70	4	-	Santos et al., 2006
8.0-8.2	850-1020	570	-	-	5-21	98-130	15-23	Coelho et al., 2006
-	68-220	0-1	-	-	0.2-21	0.9-3.8	-	Rahman & Al-Malack, 2006
8.1-8.9	510-910	-	-	-	-	30-31	-	Jou & Huang, 2003
6.5	800	-	100	100	-	8	17	Demirci et al., 1997
10	81	8	-	-	2.3	-	-	Ojuola & Onuoha, 1987
-	660-710	-	-	-	22	30	10	Serafim, 1979
-	300-600	150-250	100-300[2]	-	-	20-200		World Bank Group, 1999

[1]unitless, [2]desalter effluent

Whilst biological treatment is the most common and cost effective method for organics removal employed at oil refineries, the required treated water, which for discharge is normally between 100 and 200 mg/L COD (Ma et al., 2009; Santos et al., 2006), may be challenged by both nitrification inhibition and by the biorefractory nature of the organic fraction. A loss of nitrification can arise both from a C:N imbalance or from toxicity. In such cases where biological treatment is challenged, advanced oxidation may be necessary (Coelho et al., 2006; Saien and Nejati, 2007; Abdelwahab et al., 2009) and its implementation within the sector is becoming increasingly common.

Petrochemical effluents tend to be less challenging than refinery effluents, due to their reduced recalcitrance and water quality fluctuation. An exception is effluents containing PVA (polyvinyl alcohol) from PVC (polyvinyl chloride) manufacture, which are relatively resistant to biodegradation and thus require a high MLVSS concentration and long treatment times. This makes such effluents very conducive to treatment by MBR technology, particularly in cases where spatial restrictions exist.

3.4 Pharmaceutical wastewaters

3.4.1 Pharmaceuticals in the environment

Concerns over the occurrence and detrimental impacts of active pharmaceutical ingredients (APIs) in the environment as a whole were originally raised in the late 1990s (Daughton and Ternes, 1999), although these compounds have been detected in rivers since the mid-1990s and more recently in seawater. Various adverse impacts on aquatic life have been widely reported, most prominently the feminisation of fish, but wider impacts have also been reported, such as renal failure of vultures due to trace diclofenac (Oaks et al., 2004) and impacts on thyroid signalling in mammals (Boas et al., 2012). The biorefractory nature of many pharmaceutical compounds has meant that they tend to accumulate in the environment, albeit at very low concentrations, and it is their persistence and potential for arising in potable waters which has caused widespread concern. This is especially the case for the endocrine disrupting compound synthetic 17α-ethynylestradiol (EE2), which is not readily biodegraded and has adverse impacts even at concentrations of ng/L.

A large proportion of the API load in effluents generally derives from municipal wastewaters. A very large number of publications have appeared in learned journals on the fate and concentration of APIs in municipal wastewater and wastewater treatment plants, and CAS plants in particular. Less attention has been focused on the impact of pharmaceutical effluents (Larsson and Fick, 2009). However, effluents from pharmaceutical production are regulated and subject to limits on discharged APIs; the primary challenge to effluent treatment arises from the ancillary constituents, such as the organic solvents, fermentation broth solids and additives, which are higher in concentration and can be more recalcitrant.

3.4.2 Pharmaceutical industrial activities and effluent quality

Within the pharmaceutical industry a number of different activities exist whose generated effluents vary widely (Gupta et al., 2004). Fermentation processes (i.e. through anaerobic biological treatment) and synthetic organic chemistry are both used to produce medicines and fine chemicals. These two processes are often combined in a single manufacturing facility. Biological production (i.e. the use of live animals) is used to create vaccines and antitoxins. Finally, there are plants or operations dedicated to the production of tablets, capsules and/or solutions, though mixing, formulation and preparation.

As with many other industrial sectors (such as the food and textile industries) production varies regionally and seasonally according to demand. Wastewaters generated also vary significantly in quality according to the type of plant, with fermentation and organic chemical synthesis wastewaters generally being the highest in organic load and the most challenging to treat. Drug formulation, on the other hand, tends to produce more readily treatable wastewaters.

3.4.2.1 Fermentation

The effluent generated from fermentation is ostensibly the spent fermentation broth, the liquor remaining once the biologically-generated useful API chemicals have been extracted. Spent fermentation broth contains between 1 and 5% suspended biomass (fungi or bacteria) as well as residual solvents used for extraction. The organic fraction of the suspended matter is primarily protein and carbohydrate, with smaller quantities of fats and non-biodegradable materials. BOD concentrations can be as high as 20,000 mg/L.

3.4.2.2 Drug synthesis

Operations involved in drug synthesis include chemical reaction, solvent extraction and volatilisation, crystallisation, filtration and drying. These procedures can all generate effluent streams which may be high in both inorganic salts and COD from solvents, and chemical reaction intermediates and residual reactant. Concentrations of alcohols can be in the g/L range, and those of both chlorinated solvents and aromatic compounds (such as benzene and toluene) at several hundred mg/L. The organic content of these wastewaters thus often has significant recalcitrance, and is sometimes inhibitory to aerobic processing. Whilst more effective treatment can be provided through segregation of the different waste streams (on the basis of salinity, pH and/or organic content), this is not often carried out – particularly in older plants. By the same token, recovery of solvents is rarely viewed as cost effective since reuse for the process is not regarded as best practice (Martz, 2012).

3.4.2.3 Biological production

The use of animals in producing antitoxins, serums, vaccines and antigens leads to wastewaters containing organic matter derived from animals (manure, organs, fats, blood, etc) as well as biological culture, solvents, herbicides and other process-related components. There is therefore a pathogenic bacteria content of these wastes along with the high and variable BOD concentration (normally in the g/L range) and recalcitrant organics deriving primarily from solvents. Biorefractory organic matter is at relatively low concentrations compared with other effluent types, and these effluent streams tend to have BOD/COD ratios above 0.6.

3.4.2.4 Drug formulation

Apart from the active ingredient itself, drug formulation processes employ a range of compounds in pelletising, encapsulating and packaging a drug. Dissolved organic substances arising from the formulation itself may include sugars and syrups, alcohols, glycerin, gelatin, stabilisers and synthetic flavours. Suspended materials may include talc and diatomaceous earths. Drug formulation wastewaters tend to be slightly acidic with moderate toxicity and BOD concentrations of up to ~2,000 mg/L. However, the load from the factory process at such plants is often lower than that from the sanitary wastewater due to the labour-intensive nature of such plants.

3.4.3 Wastewater treatment

Whereas the fate of pharmaceuticals and other emerging substances of concern arising in municipal waters has received a great deal of attention since the turn of the millennium, with many reviews concerning their fate and removal (a recent example being Rivera-Utrilla et al., 2013), articles in learned journals concerning the treatment of pharmaceutical industry effluents are scarce. Reviews published since 2008 appear to have been largely limited to a review of antibiotic wastewater treatment (Guo et al., 2013), of best practices from a pharmaceutical industry perspective (Martz, 2012) and of physicochemical treatment options (Deegan et al., 2012), although a more exhaustive review was previously published as part of a reference book (Gupta et al., 2004).

Treatment of pharmaceutical wastewater draws parallels with both food industry wastewaters and in particular fermentation effluents (Section 3.2.1) and landfill leachate (Section 3.7), since fermentation is used for pharmaceuticals production and landfill leachate similarly contains high levels of biorefractory organic compounds. In general, the biological treatability of pharmaceutical wastewaters decreases with decreasing BOD/COD ratio, API concentration, and the concentration of the raw materials, intermediates and solvents. Moreover, above a COD concentration of ~4,000 mg/L CAS may be considered unsuitable (Suman Raj and Anjaneyulu, 2005). Otherwise, aerobic treatment is made viable through employing long HRTs and SRTs, which are made more accessible by the use of MBRs.

According to an extensive 2012 review (Lin et al., 2012), HRTs reported for MBR trials conducted at bench and pilot scale on analogue and actual pharmaceutical wastewaters varied from 1 to >6 days, with SRTs of 26-100 days. In all cases reported removal of COD exceeded that of the target pharmaceutical(s), with COD removal normally exceeding 90%.

3.5 Pulp and paper industry

The mean water demand of the pulp and paper (P&P) industry is currently around 55 m^3/te product. Whilst very significant, in common with many other water-intensive industrial processes, technical advances combined with increasingly stringent environmental legislation have led to substantial reductions in the water demand. In the case of the P&P sector this has been associated primarily with changes in the initial digestion (or cooking) process, which consumes around 85% of the water used for producing pulp. However, there has also been increasing use of advanced water treatment technologies and strategies geared towards water reuse and resource recovery, along with anaerobic treatment for energy recovery.

3.5.1 Pulp and paper processing

The primary digestion processes generally employed in the P&P industry are categorised as chemical, semi-chemical or thermo-mechanical. The most common, wholly chemical kraft process employs caustic soda (NaOH) and sulphide (Na_2S), combined to form "white liquor", to remove the lignin and carbohydrate substances in the wood from the fibres, to generate a cellulosic pulp from which the paper is ultimately produced. This process accounts for around two thirds of the world's pulp production, being generally favoured over other processes due to its versatility (in dealing with all types of wood feed) and intensivity (high pulp strength allowing smaller reactor vessels), combined with the potential recoverability of the chemicals. An alternative chemical process uses an acidic or neutral solution of sulphurous acid (H_2SO_3) and bisulphite (HSO_3^-), sometimes generated from sulphur dioxide (SO_2), and a base salt. Sulphite pulps have less colour than kraft pulps and so can be bleached more easily, but they are less robust and their efficacy is dependent on the type of wood (generally spruce, birch, beech and aspen) and the absence of bark.

Semi-chemical processes use a combination of mechanical and chemical processing, with the chemical process used to soften the wood chips prior to mechanical treatment. The most common reagent used for this process is sodium sulphite (Na_2SO_3) under neutral pH conditions (often termed neutral sulphite semi-chemical, or NSSC). Sodium carbonate (Na_2CO_3), either with or without caustic soda (NaOH), may also be used. Such processes employ less aggressive conditions and do not generate the odour or thiol by-product associated with the use of sulphide in the kraft process. They are, on the other hand, less intensive and generate reduced pulp yields.

The thermo-mechanical process (TMP) involves steaming the raw materials under pressure for a short period, prior to and during refining. The TMP can be further modified using chemicals during the steaming stage (hence chemithermo-mechanical pulping, CTMP). Mechanical processes are generally only suitable for newsprint paper, since the process removes little of the lignin and thus the colour. On the other hand, at around 95% they convert almost twice as much wood to pulp as the chemical digestion processes.

Pulp can also be made from recycled paper, processed into pulp and deinked prior to further usage. Fibres from recycled pulp are not as strong as in newly produced pulp, and are therefore mainly used for cardboard and corrugated paper.

Digestion is preceded by debarking and chipping, which also generates a small amount of wastewater, generally heavily polluted. Digestion is followed by coarse and fine screening to remove undigested fibre bundles and knots (with washing stages between), thickening, bleaching and possibly further refining. Subsequent paper production then employs additional chemicals such as fillers, sizing chemicals (which primarily provide water resistance), coagulants, starch and coating chemicals which then contribute to the effluent pollutant load to various degrees. The requirement for bleaching depends on the type or nature of the paper product and the lignin content of the pulp, since lignin affects the colour. The degree of delignification is generally described by the Kappa value (χ) with values ranging 25-35 for softwood and 14-23 for hardwood, prior to bleaching. This equates to a lignin content of 3-5%.

The effluent from the digester associated with the cooked chips at the completion of the kraft digestion stage, or "black liquor", contains high concentrations of salt and organic matter. In addition to the organic constituents (primarily lignin and carbohydrates), the black liquor is strongly alkaline and contains salts such as sodium carbonate (Na_2CO_3), sodium sulphate (Na_2SO_4), sodium sulphite (Na_2SO_3) and sodium thiosulphate ($Na_2S_2O_3$). Both the water and chemicals are recovered from the black liquor in a three-stage cycle. The black liquor is first evaporated, to recover the excess water. The concentrated black liquor is incinerated in a recovery boiler, where white liquor and inorganic smelt from the recovery furnace is extracted. In the third step, the organic smelt dissolved in water ("green liquor") is converted into white liquor by causticising (i.e. treated with sodium hydroxide to improve the colour). The main sources of wastewater from the recovery cycle are condensate from the black liquor evaporation and filter washing water from the causticising step. Although the process has changed little generically over the course of the past two centuries, water demand has decreased by around 50% since the late 1970s whilst the BOD load discharged has decreased by almost an order of magnitude to around 1-1.5 kg per te product.

Bleaching is normally a two-stage process with (a) oxygen delignification, and (b) final bleaching. The first step uses a highly caustic solution to promote oxidative degradation of the lignin into lower molecular weight, water-soluble compounds. These organic compounds are then removed by washing prior to the pulp entering the final bleaching stage. Magnesium sulphate is often added to aid process control, along with the oxygen. The liquor from this step forms part of the black liquor. The final bleaching step is normally conducted in stages, and may make use of a number of different reagents such as chlorine-based chemicals (chlorine gas or sodium hypochlorite), oxygen, chlorine dioxide (ClO_2), chelating agents, peroxides and/or ozone in various combinations, interspersed with washing. The propensity for chlorinated by-product generation has led to limited use of chlorine-based products.

3.5.2 Water quality and treatment

P&P effluent is characterised by high levels of suspended solids, COD and BOD from the digestion process, along with chlorinated organic (and possibly toxic) products generated by the bleaching process. Actual values for the water quality determinants vary considerably between applications according to the raw material, digestion process, bleaching sequence and paper mill integration (Table 3-10). Apart from production process related emissions, spill water sometimes arises from buffer tanks and other process steps, such that the general condition of the mill also affects the wastewater quality.

Table 3-10 Typical P&P wastewater characteristics (adapted from Pokhrel and Viraraghavan, 2004)

Process	*pH*[a]	*TS*	*SS*	*BOD*	*COD*	*Colour*[b]	*Reference*
Large mills, India	11.0	5,250	1,233	983	2,350	*Black*	Srivastava et al., 1990
Small mills, India	12.3	15,120	4,890	2,628	6,145	*Dk brown*	
Digester house	11.6	52,589	23,319	13,088	38,588	17[c]	Singh et al., 1996
Combined effluent	7.6	3,318	2023	103	675	1[c]	
TMP whitewater	4.7	-	91	1,090	2,440	-	Jahren et al., 1999
TMP whitewater	4.7	-	105	1,125	2,475	-	Jahren et al., 2002
Kraft mill	8.2	8,260	3,620	-	4,112	4,668	Rohella et al., 2001
Pulping	10	1,810	256	360	-	-	Dilek & Gokcay, 1994
Kraft mill, unbleached	8.2	1,200	150	175	-	250	Nemerow & Dasgupta, 1991
Bleached pulp mill	7.5	-	1,133	1,566	2,572	4,033	Yen et al., 1996
Bleaching	2.5	2,285	216	140	-	-	Dilek & Gokcay, 1994
Pulp & paper	7.8	4,200	1,400	1,050	4,870	*Dk brown*	Mandal & Bandana, 1996
News paper de-inking	8.3	450	400	16	78	-	Vlyssides & Economides, 1997
Paper making	7.8	1,844	760	561	953	*Black*	Gupta, 1997
Paper mill	8.7	2,415	935	425	845	*Dk brown*	Dutta, 1999
Paper machine	4.5	-	503	170	723	243	Yen et al., 1996
Paper machine	8.3	-	1,032	240	-	-	Dilek & Gokcay, 1994

All concentrations in mg/L other than: [a]unitless, [b]Pt-Co units, [c]optical density

Since the wastewater produced in the P&P processes varies greatly in properties, various options exist for its treatment and accomplishing the appropriate contaminant concentration reduction. The Integrated Pollution Prevention and Control (IPPC) organisation compiled information on best available technology (BAT) for the P&P industry in a BAT reference document (BREF) in 2001, conducted on behalf of the European Commission. A new, possibly more stringent BREF, is due late 2014. An outcome of the study was the recommended water quality characteristics for the various P&P processes attainable from application of BAT (Table 3-11).

Table 3-11 Best Available Technology (BAT) wastewater characteristics from pulp mills, as identified by the European Commission (IPPC, 2001).

Parameter	*Unit*	*Bleached kraft*	*Unbleached kraft*	*Bleached sulphite*	*Non-integrated CTMP*	*Integrated mechanical*
COD	mg/L	267-460	334-400	500-546	667-1000	167-250
BOD	mg/L	10-30	14-28	25-37	34-50	17-25
TSS	mg/L	20-30	20-40	25-37	34-50	17-25
AOX	mg/L	<8,4	-	-	-	<1.0
Tot-N	mg/L	3.4-5	7-8	4-10	7-10	4-5
Tot-P	mg/L	0.34-0.6	0.66-0.8	0.5-0.9	0.34-0.5	0.34-0

Treatment of P&P effluents has generally followed the classical process of primary clarification (either sedimentation or flotation), secondary biological treatment and clarification, and tertiary treatment (Fig 3-3). For mechanical pulp production wastewaters flocculation and chemical precipitation may also be appropriate. The choice of tertiary treatment technology depends on the required end water quality, and specifically whether reuse is the end objective. Options include granular activated carbon (GAC), membrane filtration, UV disinfection and ion exchange (IEX), with GAC being the most common for the removal of residual DOC.

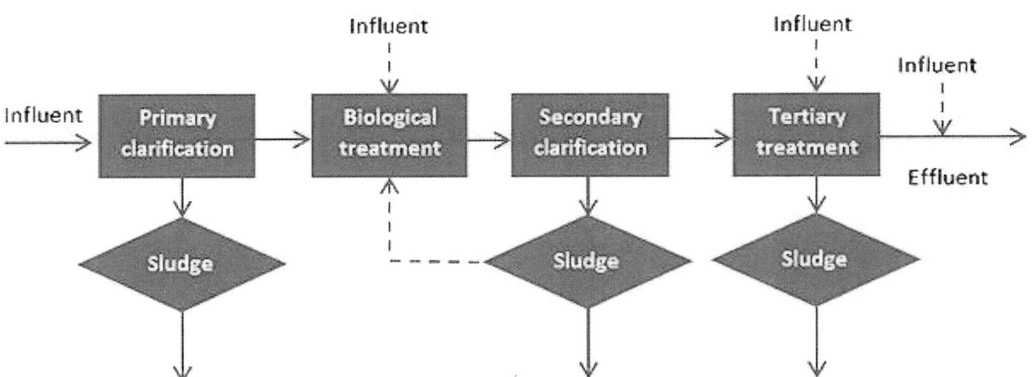

Figure 3-3 Classical P&P treatment scheme.

Effluent streams from P&P mills are commonly combined with larger streams, to be treated according to water quality, prior to entering primary, secondary and tertiary treatment. The combination of streams is also largely dependent on site characteristics, such as the location of the process steps and placement of the wastewater treatment plant.

Streams originating from the digestion (pulping) and paper production processes are characterised by high suspended solids content, mainly comprising fibres (Fig. 3-4) which are largely removed by the primary clarification stage (Fig. 3-3). Effluent from the bleaching process (or selected parts of it) may require pH adjustment prior to the biological stage. If the bleaching process also involves chlorine-based chemicals then along with organic matter the bleach effluent will contain chlorate and chlorinated organics. The condensate streams from the recovery cycle mainly contain dissolved biodegradable organic matter, such that primary treatment may not always be necessary for such streams. However, condensate generally is high in temperature, and may, along with other streams, require cooling. Other streams from the recovery cycle, such as filter wash water from the causticising, may contain phosphorus. Black liquor is strongly alkaline and can contain toxic residues such as resins from the digestion stage.

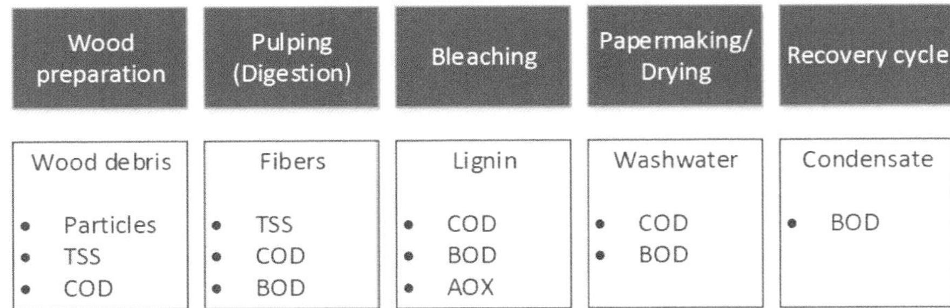

Figure 3-4 Schematic overview of the most prominent contaminant type in the P&P process steps.

3.5.3 MBRs in the P&P industry

The implementation of MBRs has become associated with P&P effluent reuse. As with other industrial sectors, implementation of water reuse in the P&P sector is dependent largely upon economics, either relating to the avoidance of punitive measures in the case of discharged pollutants or to cost benefits offered by recovered reagents and/or water. Further cost benefits may arise from reduced sludge production, if sludge generation is regarded as being particularly onerous, through operation at longer retention times in the case of the aerobic process. The high organic loads also lend these waters well to anaerobic treatment, offering both sludge solids reduction and energy recovery.

The overall high temperatures of P&P effluent make them suitable for thermophilic treatment, which may offer cost benefits due to reduced cooling costs, and simultaneously increase membrane flux. Another advantage of MBRs, as for other industries, is the comparatively small footprint compared to the conventional activated sludge process, where the clarification step (commonly sedimentation) is separate from the activated sludge step. Keeping or making a small footprint is especially important where geographical expansion areas are limited, while emission demands are increasingly stringent.

Bench- and pilot-scale studies on pulp and paper wastewater (condensate and paper mill wastewater) indicate that high levels of COD reduction are possible using MBR technology. Studies have been based on both anaerobic and aerobic conditions and in both meso- and thermo-philic environments (Table 3-12). Foul condensate seems especially conducive to MBR treatment; economic analysis has revealed the technology to offer lower operating costs than those associated with conventional stripping.

Table 3-12 MBR performance, pulp and paper (Johansson, 2012)

Wastewater	*MBR type*	*MLSS mg/L*	*Temp °C*	*Volume m^3*	*Feed COD mg/L*	*% COD rem.*	*Reference*
Foul kraft cond.	An iFS	10	37-55	0.01	10,000	97-99	Lin et al., 2009
Synthetic cond.	Ae sMT	10	55-70	0.008	1,000	>99.5	Bérubé and Hall, 2000
Foul kraft cond.	Ae iHF	3	35-55	0.004	5,000	87-97	Teixeira et al., 2005
Paper mill ww	Ae sHF	11.2	-	-	1,000	86	Galil et al., 2003
Paper mill ww	Ae sMT	-	-	-	3,000	92	Gommers et al., 2007
Paper mill ww	Ae iFS	15	-	9	1,000	89	Lerner et al., 2007
Paper mill ww	Ae iHF	8	25-34	10	600	92	Zhang et al., 2009

Whilst the use of chlorine in the pulp bleaching process may in principle cause microbial decay for biotreatment processes, this appears to be mitigated to a large degree in MBRs by the high

MLSS concentrations. Reduction of chlorate can be achieved using anaerobic MBR types. Although there are currently few full-scale MBR systems installed for P&P industrial applications, increasingly stringent emission level requirements will almost inevitably lead to increasing implementation of the technology in the future.

3.6 Textile industry effluent

Textiles can be classified according to the origin of the material. Cellulose fibres come from plant sources (such as cotton, rayon, linen, ramie, hemp and lyocell), protein fibres from animal coats (wool, angora, mohair, cashmere and silk) and synthetic fibres from polyester, nylon, spandex, acetate, acrylic and polypropylene. Most textiles are produced from cotton liners, petrochemicals and wood pulp (Ghaly et al., 2014).

As with many other industrial sectors, the textile industry consumes large amounts of water: about 80-150 L per 1 kg of fabric (Kdasi et al., 2004). It encompasses a large number of different textile processing operations (Fig. 3-5) and generates effluents of widely different characteristics with respect to specific pollutants (Table 3-13) and aggregate water quality determinants such as BOD, and TS (Table 3-14). Effluents differ in their biodegradability, toxicity, FOG content and, most notably, colour.

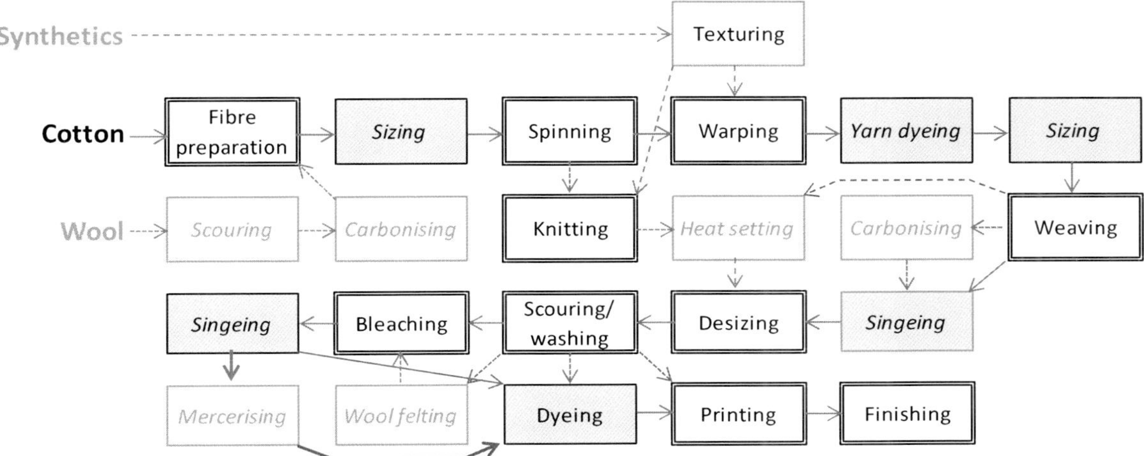

Figure 3-5 Textile processing schematic (adapted from Bisschops and Spanjers, 2003). Shaded operations may be sequenced at different places in the process. Dashed lines indicate synthetic textiles, dotted lines indicate wool, and full lines cotton.

Table 3-13 Main wastewater-generating textile processing steps (adapted from Ghaly et al., 2014)

Process stage	*Possible pollutants*	*Effluent characteristics*
Desizing	Starch, glucose, resins, fats and waxes	V. small volume, high BOD (30-50% of total)
Scouring	Caustic soda, waxes, soda ash, sodium silicate, cloth fragments	V. small volume, high pH, dark colour, high BOD (30% of total)
Bleaching	Hypochlorite, peroxide, caustic soda, acids	Small volume, high pH, low BOD (5% of total)
Mercerising	Caustic soda	Small volume, high pH, low BOD (<1% of total)
Dyeing	Dyes, mordant and reducing agents (e.g. sulphide), acetic acid, surfactants	Large volume, highly coloured, moderate BOD (6% of total)
Printing	Dyes, starch, gum oil, china clay, mordant agents, acids, metals	V. small volume, oily appearance, moderate BOD

Sizing is usually the first textile processing operation generating a significant pollutant load, the wastewater volumes being relatively low but high in organics concentration. In textile processing a "size" is a macromolecular reagent such as (modified) starch, polyvinyl alcohol, polyvinyl acetate, carboxymethyl cellulose and/or gums designed to improve the yarn (or fibre) tensile strength and abrasion resistance. The residual size concentration in the effluent and its biodegradability depends on the size, which in turn relates to the type of material, synthetic polymers generally being applied to synthetic textiles and being less biodegradable than starch-based sizes. Yarns for use as knitted fabrics are treated with lubricants (mineral, vegetable or ester-type oils) or waxes rather than sizes. Following mechanical manipulation of the yarn (spinning, warping and/or weaving into a cloth) the textile is desized (hydrolysis and removal of the size) which then generates organic matter of lower molecular weight but with a significant load.

Table 3-14 Summary of effluent characteristics (Correia et al., 1994)

Fibre	*Process*	*pH*	*BOD (mg/L)*	*TS (mg/L)*	*Water usage (L/kg)*
Cotton	Desizing	---	1,700-5,200	16,000-32,000	3-9
	Scouring	10-13	50-2,900	7,600-17,400	26-43
	Bleaching	8.5-9.6	90-1,700	2,300-14,400	3-124
	Mercerising	5.5-9.5	45-65	600-1,900	232-308
	Dyeing	5-10	11-1,800	500-14,100	8-300
	Scouring	9-14	30,000-40,000	1,129-64,448	46-100
	Dyeing	4.8-8	380-2,200	3,855-8,315	16-22
Wool	Washing	7.3-10.3	4,000-11,455	4,830-19,267	334-835
	Neutralisation	1.9-9	28	1,241-4,830	104-131
	Bleaching	6	390	908	3-22
Nylon	Scouring	10.4	1,360	1,882	50-67
	Dyeing	8.4	368	641	17-33
	Scouring	9.7	2,190	1,874	50-67
Acrylic	Dyeing	1.5-3.7	175-2,000	833-1,968	17-33
	Final scour	7.1	668	1,191	67-83
	Scouring	---	500-800	---	25-42
Polyester	Dyeing	---	480-27,000	--	17-33
	Final scour	---	650	---	17-33
Rayon	Scouring and dyeing	8.5	2,832	3,334	17-33
	Salt bath	6.8	58	4,890	4-13
Acetate	Scouring and dyeing	9.3	2,000	1,778	33-50

Following desizing the textile is scoured and washed to remove all hydrophobic materials (oils, gum, dirt, etc), as well as impurities such as metals and hardness, and leave a hydrophilic surface for the downstream washing and dyeing processes. Since caustic soda, surfactants and chelating agents are all used in the scouring process the effluent generated is alkaline and contains COD. Scouring is followed by washing and then bleaching, to remove residual colour. Bleaching of natural textiles generally employs oxidative reagents such as hypochlorite or peroxide, whereas for synthetic fibres like polyamides, polyacrylics and polyacetates reductive bleaching with hydrosulphite is used. The textile then most often undergoes dyeing and/or printing, though additional process steps may also be employed.

Process steps specific to certain textiles include mercerisation and carbonisation. Mercerisation is performed on pure cotton fabrics, which are treated by a concentrated caustic bath and a final acid wash in order to neutralise them. Its purpose is to provide lustre and increase dye affinity ("fastness") and tensile strength. Mercerisation wastewaters have low BOD and total solids levels but are highly alkaline before neutralisation. Carbonisation applies to pure woollen items, and is used to remove traces of vegetable matter. It can be carried out either in conjunction with raw scouring or at the fabric stage, depending on the level of impurities and the end use of the wool, and entails soaking in dilute sulphuric acid followed by neutralisation with sodium carbonate. Finally, a finishing step may be applied to provide a specific property such as soil repellence, waterproofing or wrinkle resistance.

Dyeing may employ a number of different dye types, with the dyes broadly classified according to their chemistry. The most commonly used dyes are soluble, including the acid, mordant, metal-complex, direct, basic, and reactive dyes: all are anionic in nature, other than the basic dyes which are cationic. About 60-70% of dyes used for dyeing textiles are azo dyes, i.e. they contain the N≡N chromophore group. They are xenobiotic compounds that are generally recalcitrant to aerobic degradation. Whilst chemical or biochemical (i.e. anaerobic) reduction can decolourise the water by attacking the azo group (Bonakdarpour et al., 2011), such processes do not substantially affect the COD concentration associated with the dye. However, such pre-treatment appears to assist downstream aerobic treatment (dos Santos et al., 2007).

Almost all dye processes leave residual colour in the effluent, the proportion of unfixed dye varying between almost negligible amounts for pigments to close to 50% for reactive dyes. This is problematic due to their high resistance to biodegradation (with most CAS plants removing no more than 50% synthetic colour) combined with the intensivity of the colour. The strong colour imparted by most dyes demands removal to very low concentrations prior to discharge for consents based on colour. Whilst insoluble dyes can be physically removed by a microporous membrane, such as indigo used for denim (Yigit et al., 2009), biorefractory soluble dyes are only rejected by dense membranes (RO or possibly NF).

Advanced oxidation processes (AOPs) can be employed to remove colour, through oxidation of the chromophore (such as the azo bond) in the dye molecule, but this does not generally mineralise the dye in the way that aerobic treatment does. There is therefore the concern that the organic by-products of AOPs, such as aromatic aldehydes and hydroxides, may be more toxic than the largely non-toxic original dyes despite the colour being removed (albeit only partially in some cases).

MBRs allow more intensive biological treatment over that attained with CAS through extending the SRT, although downstream treatment appears to be necessary if the required product water quality is based on colour. Evidence from MBR pilot studies suggests that, whilst the COD removal from textile effluents can exceed 90% (Lin et al., 2012), removal of soluble colour does not often exceed ~75% (Table 3-15). Evidence suggests that % removal is highly variable and dependent on the condition of the sludge (Brik et al., 2006), since the biorefractory nature of the dyes means that the main removal mechanism is adsorption. This is analogous to the reported behaviour of pharmaceutical and personal care products (Section 3.4) and metals; evidence suggests that residual metals levels in the treated water of the MBR are roughly half that of the CAS (Santos and Judd, 2010).

The MBR does, however, offer a clear advantage over a CAS when reuse is the end objective. A number of full-scale examples of MBR technology applied to wastewater reuse at textile mills indicate the use of downstream RO to remove the residual colour (Section 5.5). The MBR provides an effluent of low SDI suitable for downstream purification by RO or NF with no further treatment required other than cartridge filtration.

Table 3-15 Examples of published COD and colour removal data from dyehouse effluents by MBRs

		Inlet concn		*Outlet*	*% removal*		
Feed	*Technology*	*COD*	*Colour*	*Colour*	*COD*	*Colour*	*Ref*
Mill WW	AnR - Ae iMBR	196	50 DT	20 DT	80	60	Zheng & Liu, 2006
Denim mill WW, insoluble dye	Ae iMBR	1,141	2,447 Pt-Co	53 Pt-Co	97	98	Yigit et al., 2009
Analog WW, reactive dyes	Ae iMBR	2,540	-	-	93	60	Deowan et al., 2013
Polyester finishing factory	Ae sMBR	3,350	31-60 SAC	-	74-90	40-99	Brik et al., 2006

SAC spectral absorption coefficient ; Pt-Co colour on Pt-Co scale; DT dilution times: factorial dilution to sustain a target colour

3.7 Landfill leachate

3.7.1 Leachate characteristics

Landfill leachate is generated from liquids which either form part of the waste entering the landfill or from rainwater. It contains a wide variety of dissolved and suspended organic and inorganic compounds, with the organic fraction characteristics dependent on the feed composition (which relates to the type of waste delivered), the climate, the location (or "cell") within the landfill site, and leachate age. A key property of the leachate is its biodegradability (as reflected by the BOD/COD ratio), which tends to decrease with age as the waste decomposes.

Initially, aerobic decomposition takes place over the course of a few days or weeks, whereby oxygen becomes depleted and carbon dioxide is generated, causing a decline in the pH. Once oxygen is absent anaerobic and facultative organisms (acetogenic bacteria) develop. These break down cellulose and other biodegradable material to produce simple soluble compounds such as volatile fatty acids and ammonia. This stage may last several years, the leachate produced in this period being characterised by relatively high BOD/COD ratios and the presence of dissolved heavy metals which are solublised at low pH levels (Table 3-16). As time progresses slower-growing methanogenic bacteria develop which assimilate the organic compounds released during the preceding phase, generating a mixture of carbon dioxide and methane gases until the waste is mostly decomposed. When the process is over, atmospheric oxygen diffuses back to the upper layers of the landfill site, providing environmental conditions which allow the cycle to be repeated.

Leachates generated during the initial aerobic and acetogenic stages are normally referred to as "young" leachates. Those produced in the latter methanogenic stages are known as "old" or "stabilised" and, in contrast to young landfill leachates, tend to have pH values predominantly above 7 and decreased levels of BOD, COD, volatile fatty acids and metals. The concentration of ammoniacal nitrogen, depleted only very slowly, is similar to that for young leachates. The decreased biodegradability of old leachates is indicated by BOD/COD ratios generally in the range 0.05–0.2 but predominantly below 0.1 for leachates more than ~10 years old. Metals and other inorganic substances such as sulphate and chlorides continue to dissolve and leach from the landfill over many years, the prevailing anaerobic conditions reducing sulphate to sulphide. Metal sulphides are then eventually formed and precipitated back into the solid phase, maintaining low concentration values of sulphate and metals in old stabilised matrices.

Table 3-16 Leachate composition (adapted from Foo and Hameed, 2009)

Parameter	*Young*	*Intermediate*	*Old*
Age, years	<5	5-10	>10
pH	<6.5	6.5-7.5	>7.5
BOD/COD	0.5-1	0.1-0.5	<0.1
COD, mg/L	>10,000	4,000-10,000	<4,000
N-NH_4, mg/L[a]	<400	-	>400
Heavy metals	Low-medum	Low	Low
Biodegradability	High	Medium	Low

[a]can increase or decrease with age

3.7.2 Conventional landfill leachate treatment

Simple conventional aerobic suspended growth biological processes such as CAS, aerated lagoons and SBRs have all been successfully employed for treating young leachate, sometimes combined with UASB technology as pre-treatment (Sun et al., 2010). Constructed wetlands have also been implemented in some instances (Sim et al., 2013). However, these technologies are significantly challenged by old leachates due to the presence of biorefractory and toxic contaminants as well as the variable loads arising. This results in the requirement of multi-stage operations which may encompass both biological aerobic and anaerobic technologies alongside physicochemical techniques. The latter encompass physical separation by filtration or adsorption (Li et al., 2010), ammonia absorption or stripping (Hasar et al., 2009; Ferraz et al., 2013), chemical dosing for precipitation or coagulation (Hasar et al., 2009; Li et al., 2010) and/or oxidation (Zhou et al., 2010; Kilic et al., 2014), with increasing focus on electro-oxidation (Bashir et al., 2010). Consequently, the plant footprint incurred can be significant, and may still only provide modest COD removal unless aggressive oxidative conditions are employed.

A review of biological treatment of landfill leachate (Kurniawan et al., 2010) revealed a number of trends:

1. COD removal by conventional aerobic (CAS, SBR) and anaerobic (UASB) suspended growth technologies generally tends to decrease with HRT and, perhaps less demonstrably, with BOD/COD ratio as well as with COD concentration. Reported removals range from 40-98%, with the majority being over 70%. There is no apparent impact of technology: the range of reported removals is similar for CAS, SBRs, aerated lagoons and UASBs.
2. HRTs range from 0.5 to as long as 20 days, with 2-3 days being the most common.
3. For aerobic processes nitrification of over 99% is achievable provided (a) it is not carbon limited, (b) it is not inhibited by the presence of high levels of toxic metals, and (c) the nitrogen loading rates are not excessive (<0.7 kg of NH_3–N/(m^3d)), since this may also inhibit nitrification. A shortage of biodegradable carbon can lead to a pH decrease and nitrification inhibition, demanding dosing with assimilable organic carbon (e.g. methanol or acetate) if nitrification/denitrification is required.
4. Consistently high COD removal (>98%) generally demands downstream membrane-based desalination processes (RO or NF).

Extended HRTs become particularly important during the winter when temperatures are low and biological activity suppressed. Additionally, factors such as concentration of specific organic and inorganic substances, irregular production of leachate, variability of the biodegradable fraction, and variable levels of phosphorus must also be accounted for in designing biotreatment plants. Pre-treatment may also be considered if toxic substances are present (e.g. chlorinated compounds, cyanides and phenols).

A review of more recent literature reveals a generally increasing interest in integrated treatment involving, in particular, chemical oxidation combined with aerobic treatment (Cortez et al., 2011, Anfruns et al., 2013). Experimental studies of the application of chemical oxidation as pre-treatment to increase biodegradability of biorefractory organic matter have been reported for more than 20 years (Scott and Ollis, 1995). However, evidence suggests that in many cases organic matter of low biodegradability is also less readily chemically oxidised.

3.7.3 Landfill leachate treatment using MBRs

Membrane bioreactors have been implemented for the treatment of old landfill leachate since the early 1990s. They offer the benefit of both enhanced biotreatment, providing comparable COD removal to the conventional aerobic suspended growth processes at reduced HRTs (Alvarez-Vazquez et al., 2004, Fig. 3-6), and an effluent of low SDI suitable for purification by RO/NF with no further requirement for post-treatment. A few landfill leachate MBR-based technologies thus include downstream NF or RO membranes integrated into the process, such as the WEHRLE *BIOMEMBRAT® plus* process (Fig. 3-7). Whilst such post-treatment provides reliably high-quality effluent, a concentrate stream is generated as a result which demands further treatment. In the case of the WEHRLE process this stream is treated by AC and returned to the anoxic zone of the bioreactor.

A summary of MBR installations for landfill leachate treatment (Table 3-17) indicates widely varying COD removals according to, as with conventional processes, the age of the leachate (and specifically the biodegradability of the organic content) and the HRT. Studies suggest that anaerobic MBRs are as effective as aerobic technologies for removing COD, with ~90% removal at a 2-7 d HRT for young leachate (Zayen et al., 2010; Dacanal and Beal, 2010) but demand downstream treatment to remove residual ammonia.

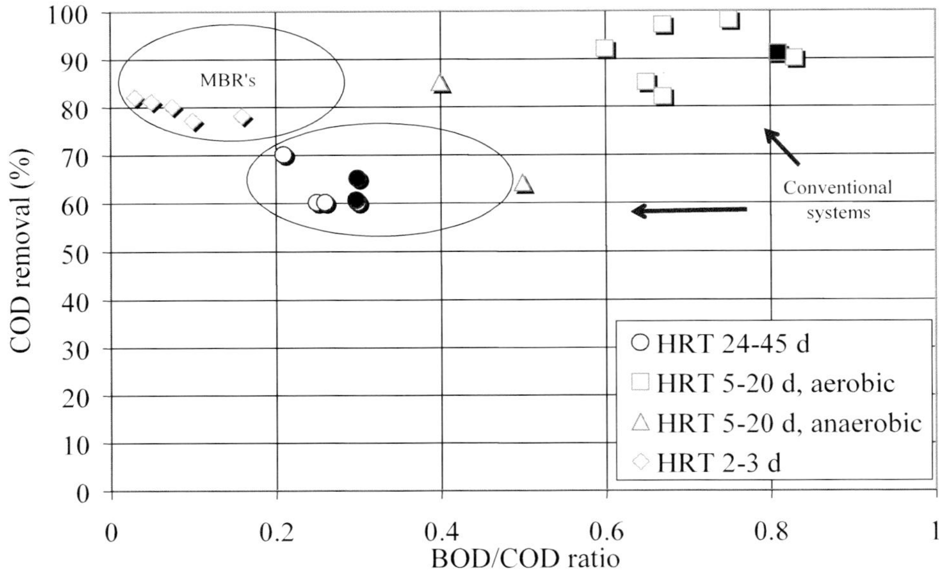

Figure 3-6 COD removal vs. leachate BOD/COD ratio at different HRTs, full-scale plant; filled data points refer to two-stage processes, encircled data are comparable (Alvarez-Vazquez et al., 2004).

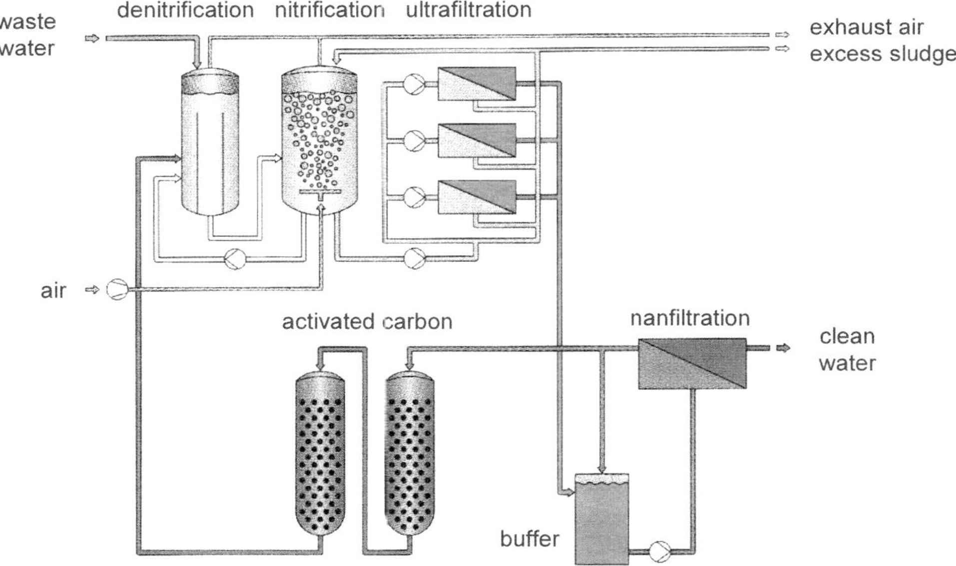

Figure 3-7 The WEHRLE *BIOMEMBRAT®* *plus* process.

Landfill leachate sludge is perhaps one of the least filterable of all those generated from MBR treatment of industrial effluents. Whereas treatment of a food processing effluent with an sMBR can yield fluxes of over 150 LMH and permeabilities in excess of 40 LMH, the application of the same system to a mature landfill leachate can yield corresponding flux or permeability values that are 2-3 times lower (Section 5.9.1).

Table 3-17 MBR performance for landfill leachate treatment, pilot plant and full-scale studies (adapted from Ahmed and Lan, 2012)

Location	*Process confign.*	*Feed, mg/L* COD	*Feed, mg/L* N-NH4	*BOD/COD age*	*HRT d*	*% removal* COD	*% removal* N-NH4	*Reference*
Thailand	Ax - Ae iHFMBR	2,600-7,300	220-1,800	*M+Y*	0.5	60-78	80-97	Chiemchaisri et al., 2011
Tianjin	An-A-LsMTMBR	4,800-6,700	820-960	*Y*		87	~100	Wang et al., 2010
Alsdorf-Warden	semi-PF sMBR	2,200	1,200	<0.05	3-7	30	90-99	Svojitka et al., 2009
Zagreb	iHFMBR	1,400-2,800	-	0.46	0.33	23	-	Jakopović et al., 2008
Beijing	AF - A-L HFMBR	10,000	-	0.71	9.5	89	-	Xu et al., 2008
Dorset	3 x Ae, sMT-MBR	5,000	2,000	0.05	-	76	~100	Robinson, 2007
Finland	SBR-iMBR	2,200	210	*Y*	2-5	>80	>97	Laitinen et al., 2006
Beijing	A-L HFMBR	4,200-16,000	-	-		70-96	-	Chen et al., 2006
Xia Ping	Ax-Ae iMBR	1,500	500	-	1.8-13	75	80-99	An et al., 2006
Penzberg	Ax-Ae MBR - AC	140-2,000	120	0.2	-	65	97	Schwarzenbeck et al., 2004
Chung-Nam	Ax-Ae iHFMBR	400-1,500	200-1,400	*M*	-	38	50-70	Ahn et al., 2002

Ae aerobic, AC activated carbon, AF anaerobic filter, A-L air-lift, Ax anoxic, iHF immersed hollow fibre, *M* mature, PF plug flow, SBR sequencing batch reactor, sMT sidestream multi-tube, *Y* young,

3.8 Ship effluents

3.8.1 Legislation

Discharge restrictions are legislated according to MARPOL (Marine Pollution Convention), the IMO (International Maritime Organisation) convention for the prevention of pollution from ships and one of the most important international marine environmental conventions. It covers prevention and minimisation of pollution from ships to the marine environment, and was signed in November 1973 prior to adoption in 1978 and introduction in October 1983 (MARPOL 73/78). The convention has six Annexes, of which Annex IV pertains to sewage and the other five to various forms of marine pollution such as oil (Annex I), bulk noxious liquids (Annex II), packaged harmful substances (Annex III), garbage (Annex V) and atmospheric discharges (Annex VI). This convention and its specifications were developed under the auspices of the Marine Environment Protection Committee (MEPC).

Sewage discharges from ships are governed by Annex IV of MARPOL, as well as other diversified regional legislation such as the US regulation USCG 33CFR 159. In the marine industry, sewage, or black water, refers to discharges from the toilets and urinals and the onboard hospital; grey water refers to discharges from showers, wash basins, laundries, and galley operations. The MARPOL convention only regulates black water, while grey water, as at July 2014, is unregulated (although subject to increasing scrutiny and subject to regulations in US waters). Bilge water, which contains high oil concentrations, rust and detergents, is regulated under MARPOL Annex I and is therefore usually treated separately by equipment subjected to a different type approval specification. For this reason, there are currently no commercial systems treating combined bilge and black/grey water.

The recommendation on International Effluent Standards and Guidelines for Performance Tests for Sewage Treatment Plants (resolution MEPC.2 (VI)) was adopted in December 1976. This resolution stipulated a discharged effluent BOD_5 and TSS of <50 mg/L and a faecal bacteria count below 250/100 mL (Table 3-18). After almost 30 years, in October 2006, this resolution was superceded by MEPC.159(55) and finally became law in January 2010. The more stringent MEPC.159(55) resolution demands a BOD_5 of <25 mg/L, a TSS of <35mg/L, and faecal bacteria below 100/100 mL, along with a COD of <125 mg/L, a pH between 6 and 8.5, and <0.5 mg/L residual chlorine. This regulation thus reduced the previous permissible sewage discharge levels of BOD by half, suspended solids by more than half, and faecal coliform levels by more than 70%, as well as imposing a very challenging residual chlorine level. As of May 2012, 129 countries, representing 87% of the world's shipping tonnage are parties to Annex IV of MARPOL 73/78, requiring all signatory flagged vessels to comply with the sewage discharge limits of resolution MEPC.159(55). In order to prevent a widespread practice of using sea water as dilution water to aid compliance performances, the Committee introduced yet another resolution, MEPC.227(64), due to come into force in January 2016. It incorporates a Dilution Compensation Factor so that any dilution water used during a type approval testing process is accounted for.

Table 3-18 Legislation governing wastewater discharges from ships

Parameter	*IMO MARPOL MEPC.2 (VI) 02/10/83*	*IMO MARPOL MEPC 159(55) 01/01/2010*	*IMO MARPOL Baltic Sea 01/01/2013*	*USCG 33CFR 159 PT1-300 20/09/1997*	*USCG Alaska 33CFR Subpart E 21/12/2000*
BOD mg/L	50	25			30[a]
TSS mg/L	50	35		150	30[a]
COD mg/L		125			
Faecal coliform /100 mL	250	100		200	20[b]
Residual chlorine mg/L		0.5			0.01[b]
pH		6~8.5			6-9[a]
N, mg/L or %removal			20 or 70%		
P, mg/L or %removal		6~8.5	1 or 80%		

[a]amended 23/06/2005; [b]amended 16/10/1984

After studies on the eutrophic marine environment and its harmful impact on the Baltic Sea, some Baltic Sea countries proposed an amendment. Resolution MEPC 200(62) was subsequently adopted on 15 July 2011, defining "Sensitive Areas" (i.e. regions more susceptible to negative environmental impact) under Annex IV; the Baltic Sea was designated as the first of such Sensitive Areas. In 2012, MEPC approved nutrient removal specifications, requiring sewage discharges from passenger ships to Sensitive Areas to have TN levels below 20 mg/L, or else >70% removal, and TP concentrations be less than 1.0 mg/L, or else >80% removal. These requirements are to be reviewed in 2014.

Since the US is not a signatory to IMO Annex IV of MARPOL 73/78, all vessels within US waters must instead comply with US discharge regulations. As with the US flagged vessels, when operating in foreign regulated waters, they must meet the requirements of Annex IV of MARPOL 73/78. IMO and US regulations thus operate in tandem.

Jurisdiction of marine pollution prevention within the US rests with the US Coast Guard. The 33 Code of Federal Regulations Part 159 (so called USCG 33CFR 159) is the regulation for overboard discharge of sewage from ships. The regulation defines three types of marine sanitation devices (MSDs):

- type I - devices producing an effluent having faecal coliform bacteria levels less than 1,000/100 mL and no visible floating solids;
- type II – devices producing an effluent having faecal coliform bacteria levels less than 200/100 mL and SS concentrations below 150 mg/L;
- type III – devices to hold the sewage on board.

The type II devices are thus those pertaining to the strictest standards set out by the US Coastal Guard (Table 3-18).

Although grey water is unregulated by MARPOL, in early 2000 Alaska State introduced tougher discharge standards (USCG 33CFR 159.309) for both black and grey water discharges from large cruise ships along with, for the first time, an independent monitoring and policing regime. Standards applicable in Alaska are: BOD_5 <30 mg/L, SS <30 mg/L, faecal coliform bacteria <20/100 mL, residual chlorine <10 microgram/L, and pH 6.0-9.0 (Table 3-18). Also, the USEPA introduced a Vessel General Permit (VGP) in 2013 which regulates grey water discharges from ships in all US waters. Grey water discharges from ships have also been regulated in the Great Lakes, as well as European inland waterways (2012/49/EU).

A voluntary sampling program conducted by the Alaska Cruise Ship Initiative in 2000 revealed shipboard sewage treatment with traditional technologies to be unable to effectively treat black water to meet the federal law USCG 33CFR 159, which required effluent concentrations not to exceed 150 mg/L and 200 faecal coliform CFUs per 100 mL (Table 3-18). The increasingly stringent regulations have subsequently precipitated a surge in the commercialisation of advanced treatment systems. These have either been based on enhanced biological systems with downstream disinfection by UV or membrane separation, or else have been based on MBR technology.

3.8.2 Wastewater management and characteristics

The sewage in ships is collected by either gravity or vacuum systems. Most use fresh water for toilet flushing to reduce corrosion, and vacuum toilets are preferred on the basis of the dramatically reduced water demand of 1.1-1.4 L per flush compared to 4.9 L for the latest water-saving, high-efficiency land-based domestic toilets (EPA, 2008). The vacuum toilet system has two direct impacts on shipboard sewage characteristics. The hydraulic loading is low at 15-25 L/day/person compared with the quantity of waste discharged by individuals in the US at 190 to 460 L/day/person on land (Tchobanoglous et al., 2004). Consequently, black water is much higher in strength (TSS, COD, BOD and other chemical constituents) than municipal wastewater. Its ammonia concentration can reach 500 to 1,000 mg $N\text{-}NH_4$/L.

Grey waters generated from the accommodation, laundries, the on-board hospital and galleys are also collected by their own dedicated piping systems and holding tanks. Grey water has 10 times the black water flow. It can have very high organic load and FOG content due to galley operations (there are typically 5-8 restaurants on a cruise ship). These different grey waters are often transferred to a series of double bottom tanks available for holding them onboard for a longer period where overboard discharges are not permitted.

The different water source characteristics support the notion of separate treatment of the streams rather than as combined black and grey water (Sun et al., 2010), though the latter option tends to be favoured by many shipboard sewage treatment suppliers. Unlike land-based sewage treatment, the design of MSDsis constrained by the confined space for installation and operation, the high strength of the waste (Table 3-19), extreme and variable hydraulic and organic shock loads, vibration, and the requirement for robustness associated with the remote location.

Differences in nutrient content between grey and black water quality become significant in regions where MEPC 200(62) applies. Thus, as marine regulations become more stringent the advantages of segregated processing of the black and grey water become more apparent. It is this which has created an opportunity for advanced wastewater treatment plants such as MBRs, for treating the high-nutrient black water waste stream since it is around one tenth of the flow but contains most of the pollutant load.

Table 3-19 Ship wastewater composition

Water quality parameter	*Black water in ships* *Vacuum toilet equipped*	*Various*	*Grey water in ships*	*Mixed black and grey water*	*Domestic wastewater*
Reference	Dokianakis et al., 2007	EPA, 2008	EPA, 2008	Sun et al., 2010	Tchobanoglous et al., 2004
TSS (mg/L)	2,100	545	318	341	120 to 400
BOD (mg/L)	650	526	354	n.a.	110 to 350
COD (mg/L)	3,097	1,140	1,000	219	250 to 800
pH	Geometric mean 7.8 (between 6.2 and 9.2)	99.5% of pH samples between 6 and 9	76.7% of pH samples between 6 and 9	Average pH 7.45	Between 6 and 9
Faec. coliform (CFU/100 mL)	n.a.	103,000,000	2,950,000 (MPN/100 mL)	n.a.	1,000 to 100,000

3.8.3 Wastewater treatment technologies on ships

Conventional sewage treatment plants on ships comprise physical/chemical and biological treatment, each having their own characteristics. Both must reduce effluent levels of organic, solid particulate and harmful pathogenic material to required standards.

Physical/chemical treatment may be based on physical methods like comminuting/ maceration, screening, and sedimentation, and chemical methods including advanced oxidation (Markle, 2006), coagulation, flocculation and disinfection. These provide a relatively small footprint because of the shorter retention time compared with biotreatment. However, reliance on chemical dosing limits their application on large ships due to the chemical handling and storage issues, along with the associated sludge production. Some commercially available physical/chemical treatment systems include *ORCA III* from Evac, which are membrane-based, and the electrochemical *OMNIPURE*™ system of Severn Trent De Nova based on electro-oxidation combined with electro-chlorination for disinfection.

Since around the early 1990s, a number of advanced biological municipal wastewater treatment technologies have been adapted and developed for marine applications (Table 3-20). These have included the SBR (Dokianakis et al., 2007), the moving bed bioreactor (MBBR, Sun et al., 2010) and, following research from the mid-1990s onwards, the MBR (Benson et al., 1999; Liu, 2009). In general, biotreatment technologies are based on either non-barrier systems (such as SBRs and MBBRs) with ultraviolet irradiation (UV) or membrane filtration for downstream disinfection, or else comprise membrane bioreactors (MBRs) for single-step biotreatment and disinfection. Some of the original MBR technologies were based on the sidestream configuration (sMBRs). However, an increasing number are now based on the immersed configuration.

Table 3-20 Commercialised marine wastewater biological treatment systems

Company	*Country*	*Product name*	*Type approved according to*	*Technology*
JOWA	Sweden	*STP 2010*	MEPC.159(55)	Submerged fixed biofilm
Gertsen & Olufsen	Denmark	*Microbac* System	MEPC.159(55)	Submerged fixed biofilm
DVZ	Germany	SKA *BIOMASTER*	MEPC.2(VI)	Submerged fixed biofilm
RWO	Germany	*WWT-LC*	MEPC.159(55)	Moving bed bio film
HAMANN	Germany	*HL-Cont PLUS* Series	MEPC.159(55)	DAF + fixed bed bioreactor + UV

Commercial treatment systems (Table 3-20) employing traditional biological processes include the *STA-C* and *ST-C* series (Hamworthy, now Wartsila), *Clarimiar* (ACO), *WWT-LC* (RWO-Veolia group), the DVZ-SKA *BIOMASTER* (DVZ) and *STP 2010* (JOWA). The latter two employ submerged fixed-bed biofilm processes, whilst RWO's *WWT-LC* technology uses an MBBR. The MBR is considered a second-generation system based on the conventional biological activated sludge system (Markle, 2006), first being used on board the Sun Princess by Hamworthy in

2000 (Liu, 2009). While MBR technologies are often higher in capital costs and demand greater operational input than the conventional technologies, they have demonstrated successful compliance performance against the most stringent discharge limits. Systems have since been introduced by many international suppliers (Table 3-21).

Table 3-21 MBR marine wastewater treatment systems

Company	*Country*	*Product name*	*Type approved according to*	*MBR configuration*
Evac	Finland	Evac MBR	MEPC. 159(55)	iFS MF
ACO	N. European	ACO-Maripur	MEPC. 159(55)	i UF
RWO-Veolia	Germany	*MEMROD*	MEPC. 159(55)	iFS MF
Martin	Germany	*siClaro® BMA®*	MEPC. 159(55)	iFS UF
DVZ	Germany	DVZ-SKA *BIOMASTER-PLUS* DVZ-JZR *BIOMASTER*	MEPC. 159(55)	s UF
Ocean Clean	Germany	OCS-COMPACT	MEPC. 159(55)	iFS MF
TRITON	Germany	TRITON MBR	MEPC. 159(55)	iFS UF
ROCHEM	Germany	*Bio-FILT®*	MEPC. 159(55)	iFS UF
Hamworthy	U.K.	MBR	MEPC. 159(55)	sMT UF

3.8.4 Outlook, ship effluent treatment

With the exception of Alaskan waters, the sewage regulations for ship discharges are often not enforced or monitored. MEPC or USCG rules demand vendors to have their equipment type approved by the classification societies, to verify and approve performance and build quality, via a series of tests under controlled conditions conducted onshore. The lack of monitoring and policing has been a serious drawback of these marine rules. While the conventional technologies are type approved, one Member State has found that the majority of the sewage treatment plants on ships did not meet the existing discharge limits (MEPC 64/23). Introduction of more stringent discharge limits has not changed this situation. On the other hand, where there is an independent monitoring regime in place, such as in Alaska where there is no type approval procedure, vendors and ship operators have demonstrated successful compliance by the use of advanced treatment systems.

3.9 Summary

Industrial waters vary widely in biodegradability and are generally subject to large changes in quality which necessitates generally between 0.5 and 2 days of equalisation. COD levels often exceed 1,000 mg/L (Table 3-22) and, whilst some industrial sectors generate effluents which tend to be very conducive to biological treatment, such as the food and beverage sector, for others treatment is far more challenging. An indication of overall biotreatability is provided by the COD ratio, which can be extremely low for specific effluents such as mature landfill leachate where BOD/COD can be below 0.1. This compares to values generally above 0.6 for the majority of food and beverage effluents.

In some cases the extreme nature of the swings in water quality associated with the upstream industrial process can be problematic despite the inclusion of a buffer tank. An example of excessive peak organic and solids loads is the impact of the desalter effluent stream in refinery wastewater treatment, which periodically discharges high dissolved/suspended organic matter and/or salinity to the WwTP. In other cases overall COD removal may be acceptable – in excess of 90% - but specific contaminants may still prove challenging because of their recalcitrance and/or impact at very low concentrations. Examples of this include dye wastes, where the colour imparted by azo dyes tends to be visible even at very low concentrations, and pharmaceutical wastewaters, for which active ingredients may be biorefractory and/or onerous to the environment or human health at similarly low residual concentrations.

Table 3-22 Summary of industrial effluent quality (adapted from Lin et al., 2012)

Food

Process	*pH*	*COD*	*BOD*	*SS*	*N-NH$_4^+$*	*TN*	*P-PO$_4^{3-}$*	*Reference*
Cheese whey	4.5-5.0	59,000-71,000	-	-	58-150	900-1,200	340-430	Farizoglu et al., 2004
Cheese whey	4.9 ± 0.27	69,000±3,300	38,000±3,000	13,500±60	-	1,100±10	500 ± 180	Saddoud et al., 2007
Dairy	7.1	5,000	2,800	3,880	0.25	17	38.6	Rajesh Banu et al., 2008
Brewery	3.5-4.5	80,000±90,000	65,000±80,000	100±150	-	110-210	90±100	Ince et al., 2000
Rice winery	4.8-5.9	30,000-35,400	16,000-19,000	310-730	-	70-140	20-30	Yu et al., 2006
Oily pet food	5-6	75,000-150,000	80,000	600-1,000	797-1,400	-	500-830	Kurian et al., 2005
Seafood processing	6.3	1,900	1,100	553	-	140	32	Sridang et al., 2008

Pulp & paper[1]

Process	*pH*	*COD*	*BOD*	*SS*	*TN*	*TP*	*TS*	*Methanol*	*Acetic acid*	*Carbohydrate*
TMP	4.2	7,210	2,800	383	12	2.3	72	25	240	2,700
CTMP	-	6,000-9,000	3,000-4,000	500	-	-	167	-	1,500	1,000
Kraft bleaching	10.1	1,100-1,800	128-184	37-74	-	-	-	40-76	0	-
Kraft foul	9.5-10.5	10,000-13,000	5,500-8,500	0	350-600	0.02-1.6	120-380	7,500-8,500	-	-
Sulfite condensate	2.5	4,000-6000	2,000-4,000	-	-	-	800-850	250	-	-
Chip wash	-	20,600	12,000	6,100	86	36	320	70	820	3,200

Textile

pH	*COD*	*BOD*	*SS*	*N-NH$_4^+$*	*P*	*Colour$_{wavelength, nm}$*	*Reference*
7.2-8.1	830-4750	115-730	-	5-18	103-118	Colour$_{669}$ = 21-140 m^{-1}	Badani et al., 2005
10.8-11.2	800-1,000	-	200-300	-	-	1,000-2,500 PtCo unit	Kim et al., 2002
9.91	1,030±67.4	170±14	180±16	23±2	2.4±0.1	Color$_{669}$ = 0.21 m^{-1}	Sen et al., 2003
2-10	50-5,000	200-300	50-500	18-39	0.3-15	>300 mg/L	Marcucci et al., 2003
10	1,150	170	150	-	-	Color$_{436-620}$=1.0-1.4 m^{-1}	Selkuk et al., 2005
9.0	1,000	300	880	57	-	-	Dantas et al., 2006

Leachate

Age	*Location*	*pH*	*COD*	*BOD*	*SS*	*TKN*	*N-NH$_4^+$*	*Reference*
Young	Hong Kong	7.0-8.3	17,000	7,300	>5,000	3,200	3,000	Lo, 1996
Young	Italy	8.2	11,00	2,300	1,700	-	5,200	Lopez et al., 2004
Medium	Germany	-	3,200	1,000	-	1,100	880	Baumgarten and Seyfried, 1996
Medium	Greece	7.9	5,400	1,000	480	1,100	940	Tatsi et al., 2003
Old	France	7.5	500	7.1	130	540	430	Trebouet et al., 1999
Old	South Korea	8.6	1,400	62	400	140	1,500	Cho et al., 2002

Pharmaceutcl.

pH	*COD*	*BOD*	*SS*	*N-NH$_4^+$*	*PO$_4^{3-}$-P*	*Cl$^-$*	*Reference*
4.2-4.5	5,000-80,000	-	900-19,000	40-320	30-120	-	Nandy et al., 2001
7-8	40,000-60,000	-	-	800-900	3-6	-	Oktem et al., 2008
6.5-10.5	850-1855	-	-	56-130	8.5-10	-	Saravanane et al., 2009
5.6	1,900	380	41	4.8	-	34,000	Melero et al., 2009
4.42	50,000	8,200	-	-	-	-	Yang et al., 2009
6.3	10,900	-	-	410	-	-	Mascolo et al., 2010

Petrolm.

Origin	*pH*	*COD*	*BOD*	*SS*	*N-NH$_4^+$*	*TP*	*Oil/Phenol*	*Reference*
Acidic petrochemical	2.5-2.7	55,000-60,000	30,000-32,000	20-300	-	100-230	360 (phenol)	Patel et al., 2002
Oil Refinery	7.8-8.8	250-610	-	110-160	56-130	<0.5	35-55 (oil)	Liu et al., 2005
Oil Refinery	-	72-300	90-190	250-950	12-20	0.8-3	20-87 (oil)	Zhidong et al., 2009
Oil Refinery	-	120	52	60	-	-	78 (oil)	Salahi et al., 2010

[1]summarised by Baipai, 2000

In such cases where, even under the biochemically intensive conditions of an MBR, aerobic degradation of some species or the bulk COD is insufficient, recourse is made to hybrid processes or, most frequently, downstream treatment with dense membrane processes (reverse osmosis or nanofiltration). The two-stage MBR-RO process appears to be becoming the treatment scheme of choice where wastewater reuse is the prime objective, particularly when spatial constraints exist. However, RO and NF serve only to separate the residual contaminants from the water, generating a concentrated waste stream which demands further management if discharge to the sewer is not an option. There is therefore increasing interest on chemical treatment, and advanced oxidation processes (AOPs) in particular, for application to some of the more challenging effluents with a significant or otherwise onerous biorefractory organics content. AOPs have been explored for both pre-treatment and post-treatment but, as yet, there appears to be no commercially-available, demonstrated MBR-AOP hybrid process technology.

References

Abdelwahab, O., Amin, N.K. and El-Ashtoukhy, E.-S.Z. (2009). Electrochemical removal of phenol from oil refinery wastewater. *J. Hazard. Mater.* **163** 711–716.

Ahmed, F.N and Lan, C.Q. (2012). Treatment of landfill leachate using membrane bioreactors: A review. *Desalination* **287** 41–54.

Ahn, W., Kang, M., Yim, S., and Choi, K. (2002). Advanced landfill leachate treatment using an integrated membrane process, *Desalination* **149** 109–114.

Al Zarooni, M. and Elshorbagy, W., (2006). Characterization and assessment of Al Ruwais refinery wastewater. *J. Hazard. Mater.* **136** 398–405.

Altas, L. and Büyükgüngör, H. (2008). Sulfide removal in petroleum refinery wastewater by chemical precipitation. *J. Hazard. Mater.* **153** 462–469.

Alvarez-Vazquez, H., Jefferson, B., Judd, S.J. (2004). Membrane bioreactors vs. conventional biological treatment of landfill leachate: a brief review, *J. Chem. Technol. Biotechnol.* **79(10)** 1043-104.

An, K.J., Tan, J.W., and Meng, L. (2006). Pilot study for the potential application of a shortcut nitrification and denitrification process in landfill leachate treatment with MBR, Water Sci. Technol. Water Supply **6** 147–154.

Anfruns, A. , Gabarró, J., Gonzalez-Olmos, R., Puig, S., Balaguer, M.D., Colprim, J. (2013). Coupling anammox and advanced oxidation-based technologies for mature landfill leachate treatment. J. Hazard. Mater. **258-259** 27-34.

Badani, Z., Ait-Amar, H., Si-Salah, A., Brik, M. and Fuchs, W. (2005). Treatment of textile waste water by membrane bioreactor and reuse. *Desalination* **185**(1-3) 411-417.

Bajpai, P. (2000). Treatment of pulp and paper mill effluents with anaerobic technology, Pira International, Randalls Road, Leatherhead, UK.

Bashir, M.J.K., Aziz, H.A., and Amr, S.A.S. (2010) An overview of electro-oxidation processes performance in stabilized landfill leachate treatment, Desalin. Water Treat., **51** (10-12) 2170-2184.

Baumgarten, G. and Seyfried, C.F. (1996). Experiences and new developments in biological pre-treatment and physical posttreatment of landfill leachate. *Water Sci. Technol.* **34**(7-8) 445-453.

Benson, J., Caplan, I. and Jacobs, R. (1999). Blackwater and graywater on U.S. navy ships: technical challenges and solutions, *Naval Engineers J.* **111**(3) 293-306.

Bérubé, P. R., and Hall, E.R. (2000). Effects of elevated operating temperatures on methanol removal kinetics from synthetic kraft pulp mill condensate using a membrane bioreactor. *Water Res.* **34**(18) 4359-4366.

Bisschops, I., and Spanjers, H. (2003). Literature review on textile wastewater characterisation, *Environ.Technol.* **24** 1399-1411.

Boas, M., Feldt-Rasmussen, U., and Main K.M. (2012). Thyroid effects of endocrine disrupting chemicals. *Mol. Cell. Endocrinol.* **355**(2) 240-8.

Bonakdarpour, B., Vyrides, I., and Stuckey, D.C. (2011). Comparison of the performance of one stage and two stage sequential anaerobic-aerobic biological processes for the treatment of reactive-azo-dye-containing synthetic wastewaters. *Internat. Biodeterioration Biodegrad.* **65** 591-599.

Brik, M., Schoeberl, P., Chamam, B., Braun, R., and Fuchs, W. (2006). Advanced treatment of textile wastewater towards reuse using a membrane bioreactor. *Proc. Biochem.* **41** 1751–1757

Chen, S., and Liu, J. (2006). Landfill leachate treatment by MBR: Performance and molecular weight distribution of organic contaminant, *Chin. Sci. Bull.* **51** 2831–2838.

Chiemchaisri, C., Chiemchaisri, W., Nindee, P., Chang, C.Y., and Yamamoto, K. (2011). Treatment performance and microbial characteristics in two-stage membrane bioreactor applied to partially stabilized leachate, *Water Sci. Technol.* **64** 1064–1072.

Cho, S.P., Hong, S.C. and Hong, S.-I. (2002). Photocatalytic degradation of the landfill leachate containing refractory matters and nitrogen compounds. *Appl. Catal. B: Environ.*, 39(2), 125-133.

Coelho, A., Castro, A.V., Dezotti, M. and Sant'Anna Jr., G.L. (2006). Treatment of petroleum refinery sourwater by advanced oxidation processes. *J. Hazard. Mater.* **137** 178–184.

Correia, V.M., Stephenson, T., and Judd, S.J. (1994). Characteristics of textile wastewaters - a review, *Environ. Technol.* **15** 917-929.

Cortez, S., Teixeira, P., Oliveira, R., Mota, M. (2011). Mature landfill leachate treatment by denitrification and ozonation. *Proc. Biochem.* **46**(1) 148-153.

Dacanal, M., and Beal, L.L. (2010). Anaerobic filter associated with microfiltration membrane treating landfill leachate, *Engenharia Sanitaria e Ambiental* **15** 11–18.

Dantas, T.L.P., Mendonça, V.P., José, H.J., Rodrigues, A.E. and Moreira, R.F.P.M. (2006). Treatment of textile wastewater by heterogeneous Fenton process using a new composite Fe_2O_3/carbon. *Chem. Eng. J.* **118**(1-2) 77-82.

Daughton, C. and Ternes T., (1999). Pharmaceuticals and personal care products in the environment: agents of subtle change. *Environ. Health Persp.* **107**(S6), 907–938.

Deegan A.M., Shaik, B., Nolan, K., Urell, K., Oelgemöller, M., Tobin, J. and Morrissey, A. (2011). Treatment options for wastewater effluents from pharmaceutical companies . *A. Int. J. Environ. Sci. Tech.* **8**(3) 649-666.

Demirci, S., Erdogan, B., Ozcimder, R. (1997). Wastewater treatment at the petroleum refinery Kirikkale Turkey using some coagulant and Turkiskh clays as coagulant aids. *Water Res.* **32** 3495–3499.

Deowan, S.A., Galiano, F., Hoinkis J., Figoli, A. and Drioli, E. (2013). Submerged membrane bioreactor (SMBR) for treatment of textile dye wastewater towards developing novel MBR process. *APCBEE Procedia* **5** 259-264.

Di Fabio S., Malamis, S., Katsou, E., Vecchiato, G., Cecchi, F. and Fatone, F. (2013). Optimization of membrane bioreactors for the treatment of petrochemical tastewater under transient conditions. *Chem. Eng. Trans.* **32** 7-11.

Dilek, F.B. and Gokcay, C.F. (1994). Treatment of effluents from hemp-based pulp and paper industry: waste characterization and physicochemical treatability. *Water Sci. Technol.* **29**(9) 161-3.

Dokianakis, S. N., Fountoulakis, M. S., Kornaros, M. and Lyberatos, G. (2007). Biological treatment of sewage from ships in a sequencing batch reactor. *Fresenius Environ. Bullet.* **16**(12) 1608-1615.

Dos Santos, A.B., Cervantes, F.J., and Van Lier, J.B. (2007). Review paper on current technologies for decolourisation of textile wastewaters: Perspectives for anaerobic technologies. *Bioresource Technol.* **98** 2369-2385

Dutta, S.K. (1999). Study of the physicochemical properties of effluent of the paper mill that affected the paddy plants. *J. Environ. Pollut.* **2-3** 181-8.

El-Naas, M.H., Al-Zuhair, S. and Alhaija, M.A. (2009). Reduction of COD in refinery wastewater through adsorption on date-pit activated carbon. *J. Hazard. Mater.* **173** 750–757.

EPA (2008). *Cruise Ship Discharge Assessment Report, December 29 2008,* EPA842-R-07-005, http://water.epa.gov/type/oceb/upload/0812cruiseshipdischargeassess.pdf. Accessed June 2014.

Fakhru'l-Razi A., Pendashteh A., Abdullah L.C., Biak D.R.A., Madaeni S.S. and Abidin Z.Z. (2009). Review of technologies for oil and gas produced water treatment, *J. Hazard. Mats.* **170** 530-551.

Farizoglu, B., Keskinler, B., Yildiz, E. and Nuhoglu, A. (2004). Cheese whey treatment performance of an aerobic jet loop membrane bioreactor. *Process Biochem.* **39**(12) 2283-2291.

Ferraz, F.M., Povinelli, J., Vieira, E.M. (2013). Ammonia removal from landfill leachate by air stripping and absorption. *Environ. Technol.* **34**(15) 2317-2326.

Foo, K.Y., and Hameed, B.H. (2009). An overview of landfill leachate treatment via activated carbon adsorption process, *J. Hazard. Mater.* **171** 54-60.

Freire D.D.C., Cammarota, M.C. and Sant'Anna G.L. (2001). Biological treatment of oil field wastewater in a sequencing batch reactor. *Environ. Technol.* **22** 1125-1135.

Galil, N.I., Sheindorf, Ch., Stahl, N., Tenenbaum, A. and Levinsky, A., (2003). Membrane bioreactors for final treatment of wastewater. *Water Sci. Technol.* **48**(8) 103-110.

Ghaly, A.E., Ananthashankar, R. Alhattab, M., and Ramakrishnan, V.V. (2014). Production, characterization and treatment of textile effluents: a critical review. *Chem Eng. Proc. Technol.* **5**(1) 1-18.

Gommers, K., De Wever, H., Brauns, E. and Peys, K., (2007). Recalcitrant COD degradation by an integrated system of ozonation and membrane bioreactor. *Water Sci. Technol.* **55**(12) 245-251.

Guo, W.-Q., Yin, R.-L., Zhou, X.-J. (2013). Current trends for biological antibiotic pharmaceutical wastewater treatment. *Adv. Mats. Res.* **726-731** 2140-2145.

Gupta, A. (1997). Pollution load of paper mill effluent and its impact on biological environment. *J. Ecotoxicol. Environ. Monit.* **7**(2) 101-12.

Gupta, S.K., Gupta, S.K. and Hung, Y-Y (2004). Treatment of pharmaceutical wastes, In: *Handbook of Industrial and Hazardous Wastes Treatment* (Wang, L.K., Hung, Y.T., Lo, H.H., and Yapijakis, C., e.d). Marcel Dekker, NY.

Halim, A.A., Aziz, H.A., Johari, M.A.M., Ariffin, K.S., and Adlan N. (2010). Ammoniacal nitrogen and COD removal from semi-aerobic landfill leachate using a composite adsorbent: Fixed bed column adsorption performance. *J. Hazard. Mater.* **175** 960-964.

Hasar, H., Unsal, S.A., Ipek, U., Karatas, S., Cinar, O., and Yaman, S. Stripping/flocculation/membrane bioreactor/reverse osmosis treatment of municipal landfill leachate. *J. Hazard. Mater.* **171** (2009) 309–317.

Ince, B.K., Ince, O., Sallis, P.J., Anderson, G.K. (2000). Inert COD production in a membrane anaerobic reactor treating brewery wastewater. *Water Res.* **34**(16) 3943-3948.

IPIECA (2010). Petroleum refining water/wastewater use and management, Operations Best Practice Series, IPIECA, London.

IPPC (2001). Reference document on best available techniques in the pulp and paper industry. European Commission, Dec 2001.

Jahren, S.J., Rintala, J.A. and Odegaard H. (1999). Evaluation of internal thermophilic biotreatment as a strategy in TMP mill closure. *Tappi J.* **82**(8)141-9.

Jahren, S.J., Rintala, J.A. and Odegaard H. (2002). Aerobic moving bed biofilm reactor treating thermomechanical pulping whitewater under thermophilic conditions. *Water Res.* **36** 1067-75.

Jakopović, H.K., Matošić, M., Muftić, M., Čurlin, M., and Mijatović, I. (2008). Treatment of landfill leachate by ozonation, ultrafiltration, nanofiltration and membrane bioreactor. *Fresenius Environ. Bull.* **17** 687–695.

Johansson, T. (2012). Application of membrane bioreactors in the pulp and paper industry, MSc thesis, Uppsala Universitet.

Jou, C.G. and Huang, G. (2003). A pilot study for oil refinery wastewater treatment using a fixed film Bioreactor. *Adv. Environ. Res.* **7** 463–469.

Kdasi, A., Idris, A., Saed, K and C, Guan. 2004. Treatment of textile wastewater by advanced oxidation processes –a review. *Global Nest Int. J.* **6**(3) 222-230.

Kilic, M.Y., Yonar, T., Mert, B.K. (2014). Landfill leachate treatment by Fenton and Fenton-Like oxidation processes. *Clean - Soil, Air, Water* **42**(5) 586-593.

Kim, B.R., Anderson, J.E., Mueller, S.A., Gaines, W.A., Szafranski, M.J., Bremmer, A.L., Yarema, G.J., Jr., Guciardo, C.D., Linden, S. and Doherty, T.E. (2006). Design and startup of a membrane-biological-reactor system at a Ford-engine plant for treating oily wastewater. *Water Environ. Res.* **78**(4) 362-371.

Kim, T.-H., Park, C., Lee, J., Shin, E.-B. and Kim, S. (2002). Pilot scale treatment of textile wastewater by combined process (fluidized biofilm process-chemical coagulation-electrochemical oxidation). *Water Res.* **36**(16) 3979-3988.

Kose B., Ozgun, H., Ersahin, M.E., Dizge, N., Koseoglu-Imer, D.Y., Atay, B., Kaya, R., Altınbas, M., Sayılı, S., Hoshan, P., Atay, D., Eren, E., Kinaci, C. and Koyuncu. I. (2012). Performance evaluation of a submerged membrane bioreactor for the treatment of brackish oil and natural gas field produced water. *Desalination* **285** 295–300.

Kurian, R., Acharya, C., Nakhla, G. and Bassi, A. (2005). Conventional and thermophilic aerobic treatability of high strength oily pet food wastewater using membrane-coupled bioreactors. *Water Res.* **39**(18) 4299-4308.

Kurniawan, T.A. Lo, W., and Chan, G.Y.S. (2006). Physico-chemical treatments for removal of recalcitrant contaminants from landfill leachate, *J. Hazard. Mater.* **129** 80–100.

Laitinen, N., Luonsi, A., and Vilen, J. (2006). Landfill leachate treatment with sequencing batch reactor and membrane bioreactor. *Desalination* **191** 86–91.

Larsson, D.G.J. and Fick, J. (2009). Transparency throughout the production chain-a way to reduce pollution from the manufacturing of pharmaceuticals. *Regul. Toxicol. Pharm.* **53**(3) 161-163.

Lerner, M., Stahl, N., and Galil, N.I. (2007). Comparative study of MBR and activated sludge in the treatment of paper mill wastewater. *Water Sci. Technol.* **55**(6) 23-29.

Li, G., Wang, W., Du, Q. (2010). Applicability of nanofiltration for the advanced treatment of landfill leachate. *J. Appl. Polym. Sci.* **116** 2343–2347.

Li, W., Hua, T., Zhou, Q., Zhang, S., Li, F. (2010). Treatment of stabilized landfill leachate by the combined process of coagulation/flocculation and powder activated carbon adsorption. *Desalination* **264**(1-2) 56-62.

Li, Y., Zhu, G., Ng, W.J., Tan, S.K. (2014). A review on removing pharmaceutical contaminants from wastewater by constructed wetlands: Design, performance and mechanism. *Sci. Total Environ.* **468-469** 908-932.

Lin H., Gao, W., Meng F., Liao, B.Q, Leung, K.T., Zhao, L., Chen, J. and Hong, H. (2012). Membrane Bioreactors for Industrial Wastewater Treatment: A Critical Review. *Crit. Revs. Environ. Sci. Technol.* **42** 677–740.

Lin, H.J., Xie, K., Mahendran, B., Bagley, D.M., Leung, K.T., Liss, S.N. and Liao, B.Q., (2009). Sludge properties and their effects on membrane fouling in submerged anaerobic membrane bioreactors (SAnMBRs). *Water Res.* **43**(15) 3827-3837.

Liu, X., Wen, J., Yuan, Q. and Zhao, X. (2005). The pilot study for oil refinery wastewater treatment using a gas-liquid-solid three-phase flow airlift loop bioreactor. *Biochem. Eng. J.* **27**(1) 40-44.

Liu, Y. (2009). Current situation of onboard wastewater treatment technologies and the applications of MBRs
船舶污水处理技术现状与 MBR 应用前景展望 *Jiangsu Ship* **26**(1) 21-23.

Lo, I.M.C. (1996). Characteristics and treatment of leachates from domestic landfills. *Environ. Int.* **22**(4) 433-442.

Lopez, A., Pagano, M., Volpe, A. and Claudio Di Pinto, A. (2004). Fenton's pre-treatment of mature landfill leachate. *Chemosphere* **54**(7) 1005-1010.

Ma, F., Guo, J.-B., Zhao, L.-J., Chang, C.-C. and Cui, D. (2009). Application of bioaugmentation to improve the activated sludge system into the contact oxidation system treating petrochemical wastewater. *Bioresour. Technol.* **100** 597–602.

Mandal, T.N. and Bandana, T.N. (1996). Studies on physicochemical and biological characteristics of pulp and paper mill effluents and its impact on human beings. *J. Freshw. Biol.* **8**(4) 191-6.

Marcucci, M., Ciabatti, I., Matteucci, A., Vernaglione, G. (2003). Membrane technologies applied to textile wastewater treatment. *Ann. N. Y. Acad. Sci.* **984** 53-64.

Markle, S.P. (2006). Design and prototype development of advanced oxidation black and gray water treatment systems, *Naval Engineers J.* **118**(3) 51-64.

Martz, M. (2012). Effective wastewater treatment in the pharmaceutical industry, *Pharmaceutical Engng.* **32**(6) 48-62.

Mascolo, G., Laera, G., Pollice, A., Cassano, D., Pinto, A., Salerno, C. and Lopez, A. (2010). Effective organics degradation from pharmaceutical wastewater by an integrated process including membrane bioreactor and ozonation. *Chemosphere* **78**(9) 1100-1109.

Melero, J.A., Martínez, F., Botas, J.A., Molina, R. and Pariente, M.I. (2009). Heterogeneous catalytic wet peroxide oxidation systems for the treatment of an industrial pharmaceutical wastewater. *Water Res.* **43**(16) 4010-4018.

Nandy, T. and Kaul, S.N. (2001). Anaerobic pre-treatment of herbal-based pharmaceutical wastewater using fixed-film reactor with recourse to energy recovery. *Water Res.* **35**(2) 351-362.

Nemerow N.L. and Dasgupta A. (1991). Industrial and hazardous waste management. New York: Van Nostrand Reinhold.

Oaks, J., Gilbert, M., Virani, M., Watson, R., Meteyer, C., Rideout, B., Shivaprasad, H., Ahmed, S., Iqbal Chaudhry, M., Arshad, M., Mahmood, S., Ali, A., Khan, A. (2004). Diclofenac residues as the cause of vulture population decline in Pakistan. *Nature* **427**(6975), 630-633.

Ojuola, E.A. and Onuoha, G.C. (1987). The effect of liquid petroleum refinery effluent on fingerlings of Sarotherodon melanotheron (Ruppel 1852) and Oreochromis niloticus (Linnaeus 1757). FAO Corporate Document Repository, Project RAF/82/009.

Oktem, Y.A., Ince, O., Sallis, P., Donnelly, T. and Ince, B.K. (2008). Anaerobic treatment of a chemical synthesis-based pharmaceutical wastewater in a hybrid upflow anaerobic sludge blanket reactor. *Bioresour. Technol.* **99**(5) 1089-1096.

Patel, H. and Madamwar, D. (2002). Effects of temperatures and organic loading rates on biomethanation of acidic petrochemical wastewater using an anaerobic upflow fixed-film reactor. *Bioresour. Technol.* **82**(1) 65-71.

Pendashteh A.R., Fakhru'l-Razi A., Chuah T.G., Dayang Radiah A.B., Madaenic S., Zurina Z.A. (2010). Biological treatment of produced water in a sequencing batch reactor by a consortium of isolated halophilic microorganisms. *Environ. Technol.* **31**(11) 1229–1239.

Pendashteh, A.R., Abdullaha, L.C., Fakhru'l-Razi, A., Madaenic, S.S., Abidin, Z.Z. and Biak, D.R.A. (2012). Evaluation of membrane bioreactor for hypersaline oily wastewater treatment. *Proc. Safety Environ. Protect.* **90** 45-55.

Qin J.-J., Oo, M.H., Tao, G. and Kekre, K.A. (2007). Feasibility study on petrochemical wastewater treatment and reuse using submerged MBR. *J. Membrane Sci.* **293**(1-2) 161-166.

Rahman M.M. and Al-Malack., M.H. (2006). Performance of a crossflow membrane bioreactor (CF-MBR) when treating refinery wastewater. *Desalination* **191**(1-3) 16-26.

Rajesh Banu, J., Anandan, S., Kaliappan, S. and Yeom, I.-T. (2008). Treatment of dairy wastewater using anaerobic and solar photocatalytic methods. *Solar Energy* **82**(9) 812-819.

Rivera-Utrilla, J., Sánchez-Polo, M., Ferro-García, M.Á., Prados-Joya, G. and Ocampo-Pérez, R. (2013). Pharmaceuticals as emerging contaminants and their removal from water. A review. *Chemosphere* **93**(7) 1268-1287.

Robinson, T. (2007). Membrane bioreactors: Nanotechnology improves landfill leachate quality. *Filtr. Sep.* **44** 38–39.

Rohella, R.S., Choudhury, S., Manthan, M. and Murthy, J.S. (2001). Removal of colour and turbidity in pulp and paper mill effluents using polyelectrolytes. *Indian J. Environ. Health* **43**(4) 159-63.

Saddoud, A., Hassairi, I. and Sayadi, S. (2007). Anaerobic membrane reactor with phase separation for the treatment of cheese whey. *Bioresour. Technol.* **98**(11) 2102-2108.

Saien, J., and Nejati, H. (2007). Enhanced photocatalytic degradation of pollutants in petroleum refinery wastewater under mild conditions. *J. Hazard. Mater.* **148** 491–495.

Salahi, A., Abbasi, M. and Mohammadi, T. (2010). Permeate flux decline during UF of oily wastewater: Experimental and modeling. *Desalination* **251**(1-3) 153-160.

Santos, A., and Judd, S. (2010). The fate of metals treated by the activated sludge process and membrane bioreactors: a brief review. *J. Env. Monitor.* **12** 110-118.

Santos, F.V., Azevedo, E.B., Sant'Anna Jr., and G.L., Dezotti, M., (2006). Photocatalysis as a tertiary treatment for petroleum refinery wastewaters. *Braz. J. Chem. Eng.* **23** 450–460.

Saravanane, R. and Sundararaman, S. (2009). Effect of loading rate and HRT on the removal of cephalosporin and their intermediates during the operation of a membrane bioreactor treating pharmaceutical wastewater. *Environ. Technol.* **30**(10) 1017-1022.

Scholz W. and Fuchs, W. (2000). Treatment of oil contaminated wastewater in a membrane bioreactor. *Water Res.* **34**(14) 3621-3629.

Schwarzenbeck, N., Leonhard, K., and Wilderer, P.A. (2004). Treatment of landfill leachate - High tech or low tech? A case study, *Water Sci. Technol.* **48** 277–284.

Scott, J. P., and Ollis, D. F. (1995) Integration of chemical and biological oxidation processes for water treatment: review and recommendations. *Environ. Prog.* **14** 88-103.

Selcuk, H. (2005). Decolorization and detoxification of textile wastewater by ozonation and coagulation processes. *Dyes Pigm.* **64**(3) 217-222.

Sen, S. and Demirer, G.N. (2003). Anaerobic treatment of real textile wastewater with a fluidized bed reactor. *Water Res.* **37**(8) 1868-1878.

Serafim, A.J. (1979). Solid Retention Time on Carbon Adsorption of Organics in Secondary Effluents From Treatment of Petroleum Refinery Waste. PhD Thesis, Texas A&M University.

Sharghi E.A., Bonakdarpour B., Roustazade P., Amoozegarb M.A., and Rabbani A.R. (2013). The biological treatment of high salinity synthetic oilfield produced water in a submerged membrane bioreactor using a halophilic bacterial consortium *J. Chem. Technol. Biotechnol.* **88** 2016–2026.

Sim, C.H., Quek, B.S., Shutes, R.B.E., Goh, K.H. (2013). Management and treatment of landfill leachate by a system of constructed wetlands and ponds in Singapore, *Water Sci. Technol.* **68**(5) 1114-1122.

Singh, R.S., Marwaha, S.S., and Khanna, P.K. (1996). Characteristics of pulp and paper mill effluents. *J. Ind. Pollut. Control* **12**(2) 163-72.

Sridang, P.C., Pottier, A., Wisniewski, C. and Grasmick, A. (2008). Performance and microbial surveying in submerged membrane bioreactor for seafood processing wastewater treatment. *J. Membrane Sci.* **317**(1-2) 43-49.

Srivastava, S.K., Bembi, R., Singh, A.K. and Sharma, A. (1990). Physicochemical studies on the characteristics and disposal problems of small and large pulp and paper mill effluents. *Indian J. Environ. Prot.* **10**(6) 438-42.

Suman Raj, D. and Anjaneyulu, Y. (2005). Evaluation of biokinetic parameters for pharmaceutical wastewaters using aerobic oxidation integrated with chemical treatment. *Proc. Biochem.* **40**(1), 165-175.

Sun, C., Leiknes, T., Weitzenböck, J. and Thorstensen, B. (2010). Development of an integrated shipboard wastewater treatment system using biofilm-MBR. *Sep. Purif. Technol.* **75**(1) 22-31.

Sun, H., Yang, Q., Peng, Y., Shi, X., Wang, S., and Zhang, S. (2010). Advanced landfill leachate treatment using a two-stage UASB-SBR system at low temperature. *J. Environ. Sci.* **22**(4) 481-485.

Svojitka, J., Wintgens, T., and Melin, T. (2009). Treatment of landfill leachate in a bench scale MBR. *Desalin. Water Treat.* **9** 136–141.

Tatsi, A.A., Zouboulis, A.I., Matis, K.A. and Samaras, P. (2003). Coagulation-flocculation pre-treatment of sanitary landfill leachates. *Chemosphere* **53**(7), 737-744.

Tchobanoglous, G., Burton, F. L. and Stensel, H. D. (2004). Wastewater engineering: treatment and reuse, 4th ed., McGraw-Hill, Boston.

Teixeira, D., Carlos, J., Passos Rezende, R., Silva, C.M. and Linardi, V.R. (2005). Biological treatment of kraft pulp mill foul condensate at high temperatures using a membrane bioreactor. *Proc. Biochem.* **40**(3-4) 1125-1129.

Tellez GT, Nirmalakhandan N and Gardea-Torresdey JL, Kinetic evaluation of a field-scale activated sludge system for removing petroleum hydrocarbons from oilfield-produced water. *Environ. Prog.* **24**(1) 96-104 (2005).

Tijani, J.O., Fatoba, O.O. and Petrik, L.F. (2013). A review of pharmaceuticals and endocrine-disrupting compounds: Sources, effects, removal, and detections. *Water Air Soil Poll.* **224**(11) 1770.

Trebouet, D., Schlumpf, J.P., Jaouen, P., Maleriat, J.P. and Quemeneur, F. (1999). Effect of operating conditions on the nanofiltration of landfill leachates: pilot-scale studies. *Environ. Technol.* **20**(6) 587 - 596.

Viero A.F., T.M. de Melo, A.P.R. Torres, N.R. Ferreira, G.L. Sant'Anna Jr., C.P. Borges, V.M.J. Santiago. 2008. The effects of long-term feeding of high organic loading in a submerged membrane bioreactor treating oil refinery wastewater, *J. Membrane Sci.* **319** 223-230.

Vlyssides, A.G. and Economides, D.G. (1997). Characterization of wastes from newspaper wash drinking process. *Fresenius Environ. Bull.* **6** 734-9.

Wang, M. F., El-Din, M. G., and Smith, D. W. (2004). Oxidation of aged raw landfill leachate with O_3 only and O_3/H_2O_2: Treatment efficiency and molecular size distribution analysis. *Ozone Sci. Eng.* **26** 287.

World Bank Group (1999). Pollution Prevention and Abatement Handbook, The World Bank (Washington).

Xu, Y., Zhou, Y., Wang, D., Chen, S., Liu, J., and Wang, Z. (2008). Occurrence and removal of organic micropollutants in the treatment of landfill leachate by combined anaerobic membrane bioreactor technology, *J. Environ. Sci.* **20** 1281–1287.

Yang, Y., Wang, P., Shi, S. and Liu, Y. (2009). Microwave enhanced Fenton-like process for the treatment of high concentration pharmaceutical wastewater. *J. Hazard. Mater.* **168**(1) 238-245.

Yen, N.T., Oanh, N.T.K, Reutergard, L.B., Wise, D.L. and Lan, L.T.T. (1996). An integrated waste survey and environmental effects of COGIDO, a bleached pulp and paper mill in Vietnam on the receiving water body. *Global Environ. Biotechnol.* **66** 349-64.

Yigit, N.O., Uzal, N., Koseoglu, H., Harman, I., Yukseler, H., Yetis, U., Civelekoglu, G., and Kitis, M (2009). Treatment of a denim producing textile industry wastewater using pilot-scale membrane bioreactor. *Desalination* **240** 143-150.

Yu, H.-Q., Zhao, Q.-B. and Tang, Y. (2006). Anaerobic treatment of winery wastewater using laboratory-scale multi- and single-fed filters at ambient temperatures. *Proc. Biochem.* **41**(12) 2477-2481.

Zayen, A., Mnif, S., Aloui, F., Fki, F., Loukil, S., Bouaziz, M. and Sayadi, S (2010). Anaerobic membrane bioreactor for the treatment of leachates from Jebel Chakir discharge in Tunisia, *J. Hazard. Mater.* **177** 918–923.

Zeng, Y., Yang, C., Zhang, J. and Pu, W. (2007). Feasibility investigation of oily wastewater treatment by combination of zinc and PAM in coagulation/flocculation. *J. Hazard. Mater.* **147**(3) 991-996.

Zhang, Y., Ma, C., Ye, F., Kong, Y., and Li, H., (2009). The treatment of wastewater of paper mill with integrated membrane process. *Desalination* **236**(1-3) 349-356.

Zheng, X., and Liu, J. (2006). Dyeing and printing wastewater treatment using a membrane bioreactor with a gravity drain. *Desalination* **190** 277–286

Zhidong L. (2010). Integrated submerged membrane bioreactor anaerobic/aerobic (ISMBRA/O) for nitrogen and phosphorus pemoval during oil refinery wastewater treatment. *Petroleum Sci. Technol.* **28** 286–293.

Zhidong, L., Na, L., Honglin, Z. and Dan, L. (2009). Study of an A/O submerged membrane bioreactor for oil refinery wastewater treatment. *Petroleum Sci. Technol.* **27**(12) 1274-1285.

Zhou, Q., Li, W., and Hua, T. (2010). Removal of organic matter from landfill leachate by advanced oxidation processes: A review. *Internat. J. Chem. Enging.* **27** 0532.

4 MBR technologies

There are two aspects to the provision of MBR technologies:

- the membrane itself, and
- the membrane filtration system.

Companies may supply membrane products as components to OEM systems builders or directly to end users, most usually with guidelines for installation and use and the usual warranties regarding membrane life. Such suppliers are not necessarily MBR systems specialists themselves, instead supplying the membrane as a commodity item. MBR OEM systems suppliers, on the other hand, provide all components associated with the operation of the membrane, i.e. the frame, membrane aerator (if submerged) and control system and, in most cases, the complete treatment technology including the biotank and screen. Suppliers predominantly serve both the municipal and industrial sectors, since the design of the membrane technology itself is very largely independent of effluent quality. Operating conditions and pre-treatment/post-treatment requirements can change significantly with feedwater characteristics, however.

The membranes are either of flat sheet (FS, Section 4.1), hollow fibre (HF, Section 4.2) or multi-tube/multi-channel (MT/MC, Section 4.3) configuration. Since no consistent terminology exists for MBR membrane components, the terms used by Judd and Judd (2011) are largely followed (Table 4-1). Accordingly, the smallest replaceable membrane component can be referred to generically as an "element". An assembly of elements is variously referred to as a module, cassette, unit, rack, stack, skid or frame by practitioners. The term "module" is widely used to define both a single HF element and an assembly of flat sheet (FS) elements. A "train", on the other hand, is normally consistently used to describe a row of membrane element assemblies in a tank.

Table 4-1 Terminology for configurations (from Judd and Judd, 2011), preferred terms **emboldened**

Component	*Flat sheet (FS)*	*Hollow Fibre (HF), rectangular*	*Hollow Fibre (HF) cylindrical; Multi-tube (MT)*
Element	**Panel**, element, cartridge	**Module**, sub-unit	**Module**
Assembly of elements	Module, **cassette**, block, unit	Rack, **cassette**, box, cartridge	Skid
	Stack, frame		

In general, the conventional immersed FS cassettes (Fig. 2-3a) comprise between 50 and 300 panels separated by 6-9 mm. The largest panels supplied within a product range are usually 0.5-1 m in width and 1-1.5 m long. Coarse-bubble aeration is provided, normally continuously, from aerators integrated with the frame and positioned beneath the cassette to air-lift sludge through the channels and generate crossflow conditions. The cassettes are usually designed to be stackable, effectively extending the length of the membrane channel and so making more efficient use of the air. Most recently, since around 2010, flat sheet-based modules formed as small blocks, sometimes termed "cinder blocks" because of their shape, have been introduced by some providers. The cinder block configuration allows the modules to be stacked within a simple metal frame to provide greater flexibility and generally higher packing densities than those attainable from the conventional "tall" panels.

Immersed HF modules (Fig. 2-2b) are normally in the region of 2 m long and, with only one or two exceptions, are not stackable. The membranes are potted and attached to permeate extracting headers at one or both ends. A degree of slack is provided to allow agitation of the fibres in the bubble stream. The aerators may be either integrated with the module or form part of the frame, and aeration may be either continuous or intermittent. More recent designs make use of an "air box", giving very vigorous intermittent aeration for a few seconds every 10-15 s.

MT modules are placed outside of the tank and can be either horizontal (Fig. 2.3c) or vertical (Fig. 2.3d) in orientation, with the latter operating through combined low-pressure pumping and the air-lifting of sludge through the membrane channels. The more established horizontally-aligned technologies rely on pumping at high flow and pressures. MC ceramic membranes, with either cylindrical or square channel geometries, have been implemented as sMBRs.

Membrane materials can be either polymeric or ceramic. The two polymers most commonly employed are polyvinylidene difluoride (PVDF) and polyether sulphone (PES). However, a small number of products exist based on high-density polyethylene (HDPE), polypropylene (PP), and polytetrafluoroethylene (PTFE). The preference for PVDF arises mainly from the combination of its chemical resistance, mechanical strength, and the relative simplicity of production of HF membranes of the required properties from the base material.

Ceramic membranes have also been employed in MBR technologies. Materials used include α-alumina (Al_2O_3), titania (TiO_2), zirconia (ZrO_2) and silicon carbide (SiC). These have been available as MT/MC configurations for a number of years. The widespread adoption of ceramic membranes has thus far been constrained by their high cost compared with polymeric materials – which themselves have tended to improve in terms of overall robustness. On the other hand, the increased number of ceramic membrane products for, specifically, immersed MBRs since around 2010 suggests that they are becoming economically competitive.

Parameters which apply to all products are the clean water permeability of the membrane (in LMH/bar) and the space occupancy of the membrane module (or cassette or skid). The latter can be expressed in terms of m^2 membrane either per m^2 projected area (area footprint, F_A) or m^3 volume occupied (volume footprint, F_V). The area footprint is of greatest practical significance in most cases, unless there is a height limit (such as on board a ship), and the value increases with stacking. F_A values are included in the following product specifications. All abbreviations and definitions are listed in the Abbreviations (page x) and Symbols sections (page xi). The following technologies are listed in alphabetical order of supplier name.

4.1 Immersed flat sheet (FS) products

4.1.1 Alfa Laval

Alfa Laval is an established Swedish heat transfer, separation and fluids handling engineering company whose environmental technology product range includes cloth filters, SBRs, all types of dewatering equipment and heat exchangers. The company's *Hollow Sheet* MBR technology was introduced in 2006, and an early plant installed in 2007 for the treatment of potato starch effluent. As of 2014 Alfa Laval *MFM* modules had been installed in more than 25 industrial MBR plants ranging from food to textile applications. The roughly square, non-rigid panel uses a 0.15 μm pore PVDF membrane in modules which can be stacked to form double- or triple-decks, the latter offering 462 m^2 of membrane area (Table 4-2). Permeate is extracted via a manifold (termed a "plenum") running along the sides of the panel, rather than via the more conventional single permeate extraction point, to promote uniform permeate flow. The membrane is ultrasonically welded to the spacer, comprising a series of open square channels, producing a composite membrane/spacer material permitting operation at very low TMPs (generally gravity fed) during the filtration cycle and chemical cleaning by back-flowing at low pressure.

4.1.2 Benenv

Benenv Co., Ltd is a Yixing, China-based environmental engineering company offering a wide range of water and wastewater treatment technologies which include sludge thickening and dewatering, chemical dosing, adsorption (including ion exchange) and membrane technologies. It provides what may be the only example of a PTFE membrane in an FS configuration (Table 4-3). The *BN* panel, available in two sizes, is otherwise based on the conventional rigid backing plate design with a single permeate extraction tube.

Table 4-2 Alfa Laval *Hollow Sheet* membrane and module specifications

Material	PVDF		
Pore size, ave (max), µm:	0.2 (0.4)		
Clean water permeability, LMH/bar:	4,000		
Model:	***MFM100***	***MFM200***	***MFM300***
Height, mm:	1600	2878	3900
Width, mm:	1062		
Length, mm:	1122		
Membrane area, m^2:	154	308	462
Packing density, m^2/m^2:	129	258	388
Membrane air scour rate, $Nm^3/(m^2.h)$:	0.48-0.54	0.24-0.3	0.18-0.24
Clean water permeability, LMH/bar:	>3,000		
Recommended TMP range, mbar:	10-60		

The three models represent the single-, double- and triple-deck module

Table 4-3 Benenv panel and module specifications

Material:	PTFE		
Pore size, µm:	0.2-0.4		
Model	Panel		Module
	BN90	***BN150***	***BN150-100/150***
Height, mm:	1030	1650	1600-2350
Width, mm:	510	510	650
Length, mm:	-	-	2650
Membrane area, m^2:	0.9	1.5	150-225
No. panels/cassette:	-	-	87-150
Packing density, m^2/m^2:	-	-	52-130
Membrane air scour rate, $Nm^3/(m^2.h)$:	0.67-0.8	0.6-0.8	-
Flux, m/d	21-29		
Recommended TMP range, mbar:	50-200		

Single-deck module specifications provided: area packing density increases with stacking.

4.1.3 Brightwater/Anua

Brightwater (part of the F.L.I. Environmental group based in Ireland) designs, supplies and commissions equipment for sewage and industrial effluent treatment. Technologies include BAFs, SAFs and MBBRs. The company's *Membright®* MBR is fitted with a 150 kDa PES membrane mounted on a rigid polypropylene (PP) support, each panel providing an area of 1.84 m^2 (Table 4-4). The module is almost square in aspect, 1,120 mm long, 1,215 mm wide and 1,450 mm high for a 50-panel unit which yields 92 m^2. The same membrane panel is used by the Bord na Móna company Anua (also from Ireland) in its *PuraM®* MBR technology.

Table 4-4 Brightwater *Membright®* and Anua *PuraM®* membrane and module specifications

Material	PES		
Pore size or MWCO	150 kDa		
	Panel	Module	
Length or height, mm:	950	1450	1450
Width, mm:	950	1215	715
Thickness or breadth, mm:	7	1120	1120
Membrane area, m^2:	1.84	92	46
No. panels/cassette:	-	50	25
Packing density, m^2/m^2:	-	68	57
Membrane air scour rate, $Nm^3/(m^2.h)$:	0.69		
Clean water permeability, LMH/bar:	150		
Recommended max TMP, mbar:	350 mbar		

4.1.4 Ceraflo

Ceraflo Pte Ltd of Singapore is one of the growing number of recent suppliers offering a ceramic FS product, based on a "cinder block" design (Table 4-5). Both pressurised (sidestream) and non-pressurised (immersed) versions of the *CS* module are offered, the permeate being collected from one side of the blocks.

Table 4-5 Ceraflo immersed and sidestream membrane block specifications

Material:	alumina	
Pore size, μm:	0.5	
Model	***CS36***	***CS46***
Height, mm:	143 (145*)	
Thickness or width, mm:	360 (428*)	
Breadth, mm:	560 (600*)	
Membrane area, m^2:	4	5
Number of plates per block:	36	46
Packing density per block*, m^2/m^2:	20	24
Recommended flux range, LMH:	30-50	
Recommended TMP range, mbar:	100-500	

Data refer to 1 block: area packing density increases with stacking up to 160 m^2/m^2 for a stack of 8 blocks; *sidestream module.

4.1.5 Ecologix

Ecologix Technologies Asia Pacific, Inc. in Taiwan provides a number of technologies relating to water and wastewater treatment, most with the pre-fix "*Eco*". They include internally and externally-fed rotary drum screens, a dissolved air filtration (DAF) technology (enhanced with lamella separators), a range of disc and tube membrane diffusers, a pipe flocculator, clarifier, and a fabric biofilm carrier. Their membrane products comprise an FS-based MBR module (the *EcoPlate™*, Table 4-6) and associated *Compact MBR* package plant.

The *EcoPlate™* (Fig. 4-1a) is based on a 0.08 μm PVDF membrane, originating from the US, supported by a PET non-woven layer. The membrane is glued and ultrasonically welded onto a conventional rigid 6 mm-thick ABS support plate fitted with a nozzle for permeate extraction. Two panel sizes of 0.86 m^2 and 1.33 m^2 membrane area are offered, placed into the cassette (Fig. 4-1b-c) with a panel spacing of 6 mm. Both single- and double-deck units are available, with up to 400 panels fitted in the case of the double-deck. The *EK* and *EW* cassettes comprise a membrane element block with a patented air scouring block developed by the company.

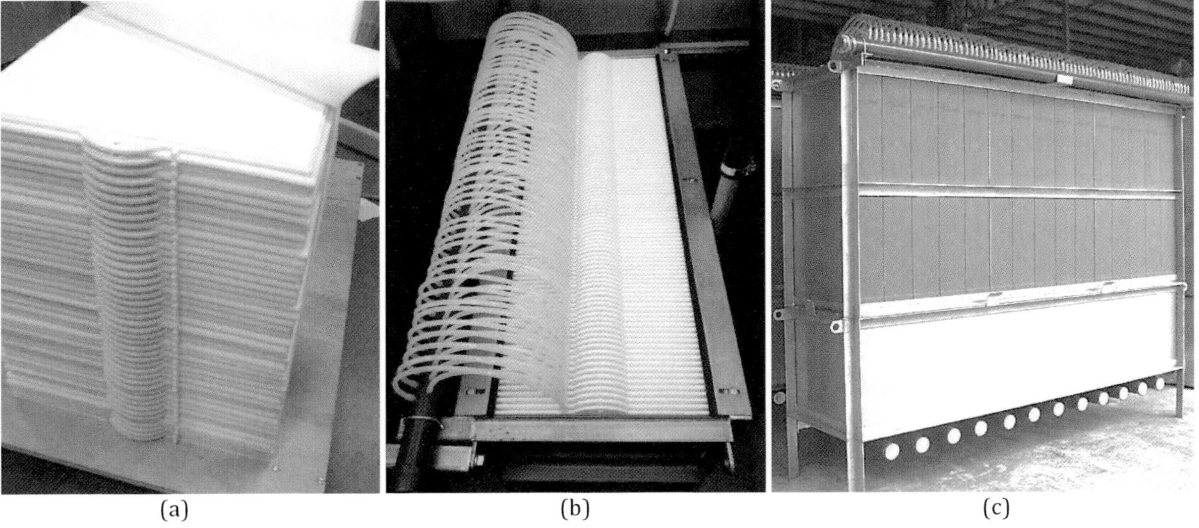

(a) (b) (c)

Figure 4-1 Ecologix technology (a) panels, and (b, c) module.

Table 4-6 Ecologix *EcoPlate™* panel and module specifications

Material Pore size, μm:	PVDF 0.08					
Model:	***EK-08***			***EW-14***		
	Panel	Module, s*	Module, d*	Panel	Module, s*	Module, d*
Length or height, mm:	1000	1830	3160	1500	2330	4160
Width, mm:	490	610	610	490	610	610
Thickness or breadth, mm:	6	940-2900	940-2900	6	940-2900	940-2900
Membrane area, m²:	0.86	51.6-172	103-344	1.33	80-266	160-532
No. panels/cassette:	-	60-200	120-400	-	60-200	120-400
Packing density, m²/m²:	-	90-97	180-194	-	140-150	209-226
Membrane air scour rate, Nm³/(m².h): Flux, LMH: Recommended max TMP, mbar	0.69 17-35 200 mbar					

s, single; d, double

As of 2014, there were over 120 Ecologix systems in operation around the world challenged with a variety of different municipal and industrial wastewaters. The latter include ~1,000 m³/d food processing effluent, ~1,400 m³/d slaughterhouse effluent, 1,500 m³/d of landfill leachate and 4,800 m³/d of petrochemical effluent.

4.1.6 Huber

Huber SE is a German company known mainly for its wastewater screening and sludge treatment equipment. The design of its MBR membrane module *VRM®* (or *Vacuum Rotation Membrane*) is almost unique in that it rotates, at a frequency of 1–2 rpm, around an axis where two coarse-bubble aerator tubes are housed. The rotary action allows all areas of the membrane to receive air scouring, efficiently removing any solids that may otherwise collect in the membrane interstices and apparently removing the requirement for chemical cleaning. The 0.038 μm pore-size PES membrane elements comprise a four-panel segment (Fig. 4-2a) of a complete 2.3 or 3.2 m diameter plate of which a hundred form the *VRM®* 20 or *VRM®* 30 module (Fig. 4-2b, Table 4-7). The 6 mm-thick panels are each fitted with a single permeate extraction tube, which feeds into a manifold running along the length of the module and into a central collecting pipe at the axis (Fig. 4-2c).

Huber was for some time the only provider of a rotating MBR membrane module. The company also produces a package plant (the *BioMem®* system) which is based on more conventional rectangular FS panels but whose livery can be customised.

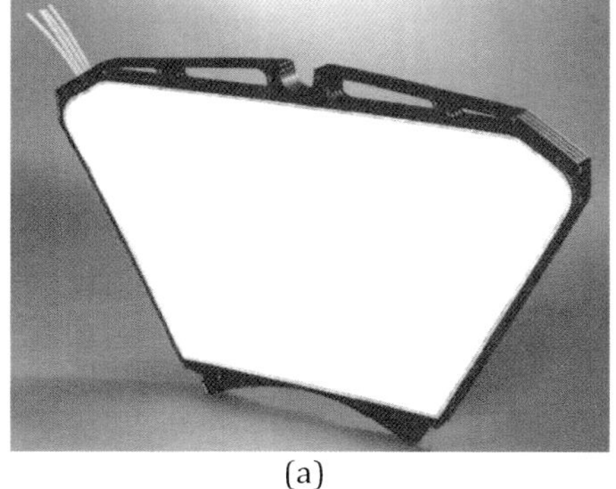

(a)

(b)

(c)

Figure 4-2 The Huber *VRM®* membrane module: (a) the four-panel element, (b) schematic, end view, (c) module, showing permeate collection tubes and manifold

4.1.7 ItN

The German company ItN Nanovation AG offers a ceramic FS product, *CFM Systems®*. It is based on a "cinder block" design, in which 35 rectangular ceramic plates are formed into a block (Fig.

4-3a), and these blocks are then stacked to form *SP* "towers" (Fig. 4-3b). Whilst the overall packing density is relatively low, the use of ceramic membranes permits the application of aggressive chemicals for cleaning along with backflushing at high pressure, such that high-flux operation may be possible (Table 4-8).

Table 4-7 Huber *VRM*® membrane and module specifications

Material:	PES	
Pore MWCO, kDa:	150	
Clean water permeability, LMH/bar	>1000	
Model	***VRM®20***	***VRM®30***
Diameter, mm:	2300	3200
Length, mm:	4000	6500
Max membrane area, m^2:	900	3480
Max. number of plates:	100	100
Packing density, m^2/m^2:	98	167
Membrane air scour rate, $Nm^3/(m^2.h)$:	0.15-0.25	
Flux, LMH:	18 cont, 33 peak	20 cont, 37 peak
Recommended max TMP, mbar:	500	

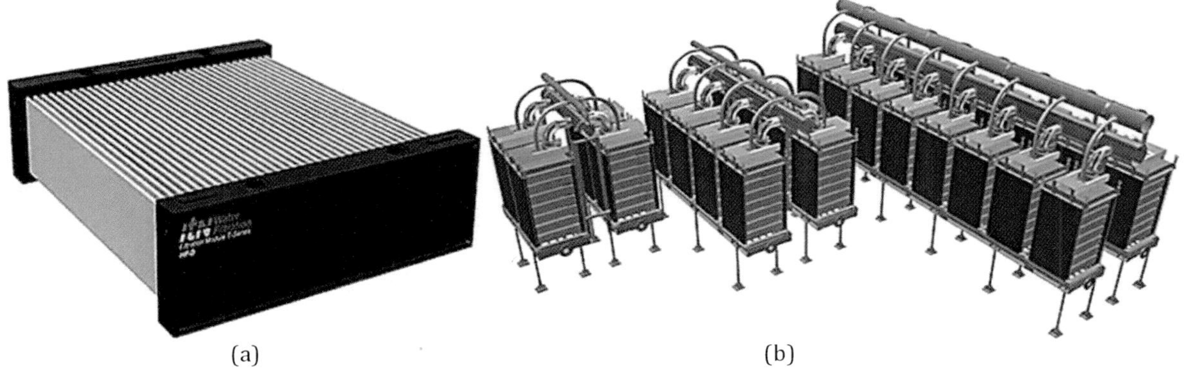

(a) (b)

Figure 4-3 The ItN system (a) membrane block, and (b) stacks (or towers)

Table 4-8 ItN *CFM Systems*® (Ceramic Flat Membrane) specifications

Material:	α-alumina/zirconia		
Pore size, μm:	0.2		
Clean water permeability, LMH/bar	6000		
	Plate	Block	Tower (***SP***)
Height, mm:	110	140	3150
Thickness or width, mm:	6.5	510	1600
Breadth, mm:	530	610	1370-5355
Membrane area, m^2:	0.12	4	192-768
Number of plates/blocks[a]:	-	35	48-192
Packing density, m^2/m^2:	-	13	87-90
Max flux, LMH:	70		
Recommended TMP range, mbar:	700		
Backwash pressure, mbar:	1500		

[a]Plates per block; blocks per stack

4.1.8 KOReD

KOReD of Korea, established in 1999, launched the *Neofil*® FS membrane panel in 2003. The company subsequently formed a joint venture with the Korean electronics giant LG Electronics in 2009, launching the Green Membrane Bioreactor (G-MBR) the following year. LG have since formed a joint venture with Hitachi (Section 4.1.24). The KOReD membrane is of PES with a pore size of 0.2 µm. The technology is based on a rigid rectangular panel, 1.0 m^2 in area and with dimensions of 1,200 x 490 x 16 mm thick, with a single permeate extraction point.

4.1.9 Kubota

The Kubota membrane panel (referred to as a cartridge) was the first MBR membrane to be commercialised, following demonstration trials in 1990. The original type *510* panel (Fig. 4-4a), 0.8 m^2 in membrane area, remains in widespread use almost 25 years after it was first developed. As at May 2014 the company reported 1,160 MLD of capacity from 2,571 municipal wastewater treatment plants, and 484 MLD of capacity from 2,262 industrial installations. At over 4,800 installations in total worldwide, there are more wastewater treatment installations based on the Kubota technology than for any other supplier.

The membrane itself is unusual, being the only MBR membrane which is permanently hydrophilicised through partial chlorination of a polyethylene (PE) material. The membrane (0.2, 0.4 µm average, maximum pore size) is supported by a robust non-woven substrate and welded on all sides to a rigid backing plate with a spacer material sandwiched between the membrane and plate. Narrow channels cut into the plate ensure even collection of permeate across the surface, and permeate is extracted from a single outlet at the top of the panel via transparent PVC tubing. Between 75 and 200 panels are then fitted into a cassette or "submerged membrane unit", SMU (Fig. 4-4b), with a manifold collecting the permeate from each panel.

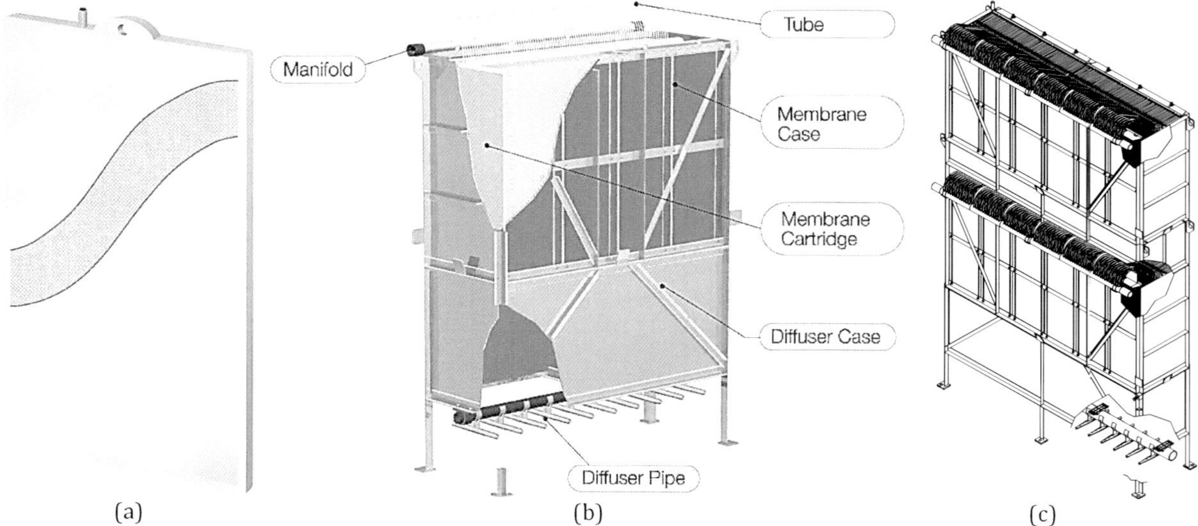

Figure 4-4 (a) Kubota panel (showing membrane, spacer and backing plate), (b) submerged membrane unit (SMU), and (c) double-deck unit

Aeration via coarse bubble aerators is applied at the base of the tank. The original loop-type aeration pipe has been largely superseded by a flushable "centapedal" aerator. Cleaning of the aerator is achieved by briefly opening an external valve connected via a manifold to the ends of the central pipe(s). This allows vigorous backflow of sludge and air back into the aerator, flushing out any solid deposits which would otherwise accumulate and dry out in the aerator.

Membrane modules are provided in single-deck (the *ES*, Fig. 4-4b) or double-deck configurations (the *EK* series Fig. 4-4c), the latter being introduced in 2002, as well as the newer *SP* "block" design (Fig. 4-5). The double-deck design reduces capital costs by doubling the membrane surface area per unit plant footprint, whilst at the same time halving the number of diffusers and cases. The operational costs are also reduced through the reduction in the specific aeration demand from 0.75 to 0.53 Nm^3/(m^2.h) (Table 4-9).

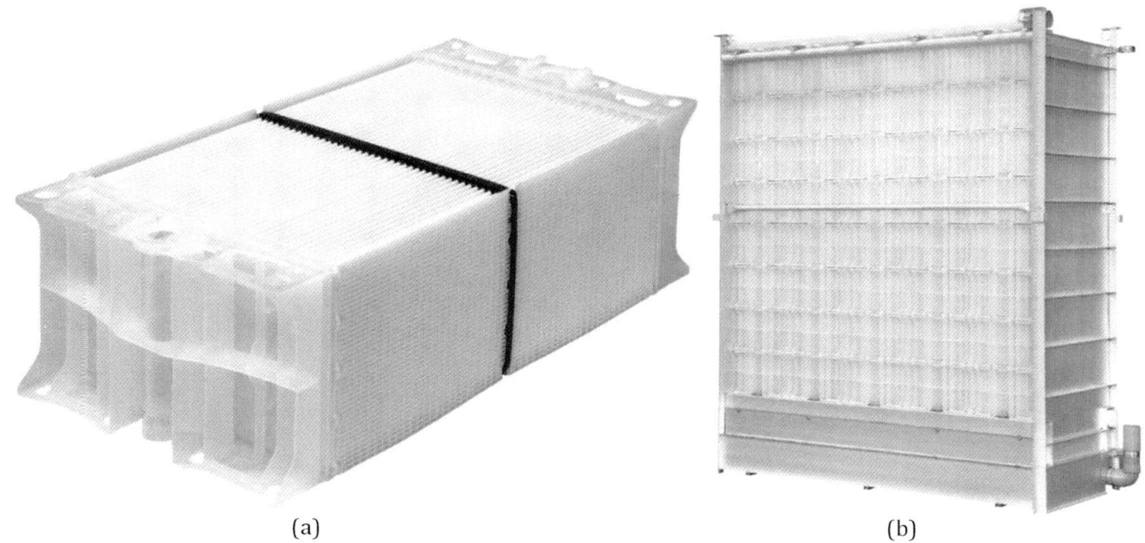

Figure 4-5 The Kubota *SP400* (a) block, and (b) stack (or SMU)

Table 4-9 Kubota membranes and modules specifications

Material:	PE (chlorinated)					
Pore size, µm:	0.2 ave, 0.4 max					
Clean water permeab., LMH/bar:	1390					
Model	***ES*** **pan.**	***ES*** **cass.**	***EK*** **cass.***	***RM*** **pan.**	***RM*** **cass.**	***RW*** **cass.***
Length or height, mm:	1020	2030	3500	1560	2490	4290
Width, mm:	490	600-620	600-620	575	575	575
Thickness or breadth, mm:	6	1140-2920	2200-2920	6	2250-2930	2250-2930
Membrane area, m^2:	0.8	60-160	240-320	1.45	218-290	435-580
No. panels/cassette:	-	75-200	300-400	-	150-200	300-400
Packing density, m^2/m^2:	-	88	177-182	-	169-172	336-344
Membrane air scour rate, Nm^3/(m^2.h):	-	0.75	0.53	-	0.42	0.29
Recommended TMP range, mbar:	50-200					
Year of commercialisation:	1990		2003	2009	2010	2009

*double-deck

Kubota introduced a longer panel, the 1.45 m^2 area *515*, in 2009. This panel has two permeate extraction nozzles to improve hydrodynamics. The *RW400* provides up to 400 panels (580 m^2 filtration area, Table 4-9). The increased panel length further reduces both the footprint and the membrane aeration demand; the recommended guide value of SAD_m for the double-deck module is 0.29 Nm^3/(m^2.h)(Fig. 4-4c). At a total height of 4.3 m, the module is around 2.1 times the height of the original *ES* module. The operating conditions, the materials and the concept design of the original and new modules are otherwise the same. The first MBR plant based on the double-deck *RW* module began operation in May 2009. Kubota launched the *RM* series, single-deck modules accommodating Type *515* membrane cartridges, in early 2010.

The most recent Kubota product, the *SP*, differs in design from the *ES* and *RM* "long panel" type products. It provides a unit membrane area of up to 400 m^2 through stacking an array of "cinder block"-type modules (Fig. 4-5), 10 m^2 in membrane area, in an array of 5 x 8 high (i.e. 40 modules). The modules are small enough to be extracted manually from one side of the unit once it is removed from the tank. This avoids the requirement of disassembling a double-deck module in order to access the lower deck.

4.1.10 Lantian Peier

The Jiangsu Lantian Peier Membrane Co., Ltd. claims to have China's largest clean production facility for fabricating FS membranes. The company offers standard 518 mm-width PVDF membrane panels at lengths of 1.16 and 1.75 m (Table 4-10) for both domestic and industrial effluents. The company thus provides what is probably the longest commercially available FS membrane panel.

Table 4-10 Lantian Peier panel specifications

Material:	PVDF		
Pore size, μm:	0.2-0.4		
Clean water permeability, LMH/bar:	-		
Model	***PEIER-80***	***PEIER-100***	***PEIER-150***
Height, mm:	1000	1162	1752
Width, mm:	490	518	518
Length, mm:	1000	1162	1752
Membrane area, m^2:	0.8	1	1.5
Membrane air scour rate, $Nm^3/(m^2.h)$:	0.72	0.72	0.48
Recommended flux range, LMH:	16-29		

4.1.11 LiqTech

LiqTech International A/S is a Danish company specialising in silicon carbide (SiC) multi-channel membranes, offered in a range of pore sizes. The FS configuration (as the *FSM* product range, Table 4-11) was introduced in 2013 and, as of 2014, it was the only company providing FS membranes of SiC. This material is arguably the most robust of the ceramic membranes offered, being highly resistant to both aggressive chemicals and organic fouling.

Table 4-11 LiqTech membrane and rack specifications

Material:	SiC				
Pore size, μm:	"Range from UF to MF"				
Model	**Module**		**Rack**		
	FSM 350	***FSM 550***			
Height, mm:	150	190	1300	2600	2600
Width, mm:	561	561	670	670	670
Length, mm:	540	566	670	670	1400
Membrane area, m^2:	3.5	5.5	22	44	88
No. modules:	-	-	4	8	16
Packing density, m^2/m^2:	12	17	49	98	94
Max flux, LMH:	710-1100				
Max TMP, mbar:	5000				

4.1.12 Martin

Martin Membrane Systems AG of Germany offers the PES-based *siClaro*® system based on small, stackable modules of 6.25 m^2 membrane area (the *FM 6* range) or 20 m^2 (*LFM 20* range) fitted with an unspecified number of membrane sheets (Table 4-12). These then form stacks of 1-16 modules per level and 1-4 levels, thereby providing flexibility in the size or space occupancy. There were around 20,000 installed membrane modules across a range of municipal and industrial applications, the latter including the food and beverage, pulp and paper, cosmetics, landfill leachate and marine industries, by mid 2014.

4.1.13 MaxFlow / A3

MMF MaxFlow Membran Filtration GmbH is a German company supplying MBR membranes (*MaxFlow*) for industrial and municipal applications. A3 Water Solutions GmbH is a sister company of MMF and builds, as a construction company, MBR wastewater treatment plants for niche, relatively low-flow applications such as army field camps, ship-board effluent treatment, food industry wastewater streams, etc. The product range of *MaxFlow* membrane modules (Table 4-13) comprises the 0.14 μm pore-size PVDF and the 150 kDa PES models. Three size ranges are offered for the two membranes as serial modules: 6, 20 and 70 m², the largest size for wastewater applications comprising 55 sheets of approx. 1.35 m² filtration area. The largest modules operate at a recommended SAD_m of 0.69 Nm³/(m².h) for a single deck and 0.2 for a triple deck, aeration being via tube aerators. The construction of the panels, which are of a composite material encompassing both the membrane and spacer, means that they are backflushable at pressures below 50 mbar.

Table 4-12 Martin *siClaro*® module specifications

Material Clean water permeability, LMH/bar: MWCO, kDa (Pore size, μm):	Organic polymer, PES - 150 (0.035)					
Model:	***FM 6123***	***FM 6143***	***FM 6163***	***FM 6144***	***FM 6164***	***LFM 20103***
Height, mm:	2058	2373	2372	2373	2688	2810
Width, mm:	2247	2247	2775	2247	2775	3103
Thickness or breadth, mm:	642	642	642	642	642	642
Membrane area, m²:	225	262.5	300	350	400	600

Table 4-13 MaxFlow membrane and module specifications

Model (*MaxFlow*):	***U06 - U70***	***M06 - M70***	
Material:	PES	PVDF	
Pore size or MWCO:	150 kDa	0.2 μm	
Size:	***U* or *M06***	***U* or *M20***	***U* or *M70***
Length or height, mm:	1070	1070	1070
Width, mm:	185	385	736
Thickness or breadth, mm:	316	466	716
Membrane area, m²:	6	20	70
Packing density, m²/m²*	103	111	133
Membrane air scour rate, Nm³/(m².h)*:	0.8	0.8	0.69
Clean water permeability, LMH/bar:	>300 (U series) >1000 (M series)		
Recommended TMP range, mbar:	20-250		
Backwash pressure, mbar:	<50		

*single-deck

4.1.14 MegaVision

Shanghai MegaVision Membrane Engineering & Technology Co., Ltd offers both HF and FS membranes, first introducing its PVDF membrane FS panel (Table 4-14) for MBR duties in 2006. The product is based on a conventional rigid panel and has been employed for various regional industrial effluent treatment applications. The latter have included a number of pharmaceutical effluent treatment plants located in Zhejiang province, in conjunction with the *Jet-Loop System*©® (of Valorsabio, Portugal) for enhanced oxygen transfer technology. As of May 2014 the total installed capacity provided by all MegaVision FS membranes (the *FMBR* and *RMBR* ranges) was around 255 MLD, 4% of this relating to industrial installations.

4.1.15 Meiden

Meiden Singapore Pte Ltd offers an FS ceramic (alumina-based) MF membrane of the conventional "long panel" geometry provided as a single or double deck (Table 4-15) in the company's *TM* range. The technology was selected for demonstration testing as an aerobic polishing MBR unit installed downstream of a UASB at the Jurong Water Reclamation Plant in Singapore, the trials beginning in 2014. The 4.5 MLD capacity demonstration plant is challenged with high-strength industrial effluent and the treated water to be considered for non-potable reuse.

Table 4-14 Shanghai MegaVision panel and module specifications

Material Pore size, µm:	PVDF 0.05-0.65					
Model:	***FMBR-1.0-100***		***RMBR-1.2-100***		***RMBR-1.5-100***	
	Panel	Module (single-deck)	Panel	Module (Single-deck)	Panel	Module (Single-deck)
Length or height, mm:	930	1770	1220	2538	1520	2538
Width, mm:	610	715	520	915	520	915
Thickness or breadth, mm:	16*	1860-2700	16	2090	16*	2400
Membrane area, m²:	1	100-150	1.2	120	1.5	150
No. panels/cassette:	-	100-150		100	-	100
Packing density, m²/m²:	-	81-84			-	
Membrane air scour rate, Nm³/(m².h): Flux, LMH:	0.75 16-25					

*includes both membrane and channel. Module size dependent on tank size and capacity.

Table 4-15 Meiden membrane and module specifications

Material Clean water permeability, LMH/bar: Pore size, µm:	alumina 1700 @ 1 bar 0.1		
Model prefix: "*CH250-1000*"	***E01K-5NA***	***TM100-U1DJ***	***TM100-U2DJ***
	Panel	Single-deck	Double-deck
Length or height, mm:	1046	1750	3300
Width, mm:	261	2100	2100
Thickness or breadth, mm:	6	800	800
Membrane area, m²:	0.5	100	200
No. panels per cassette:	-	200	400
Packing density, m²/m²:	-	60	119

4.1.16 MICRODYN-NADIR

The MICRODYN-NADIR *BIO-CEL*® MBR membrane technology (Table 4-16), originally introduced in 2005, is based on a composite laminated material. A 0.04 µm PES membrane is permanently affixed to each side of a macroporous polyester separator to form a flexible, thin (2 mm) FS panel. The membrane sheets, which are backflushable, are formed into cassettes (the *BC* product range) which are mounted to the sides of the frame. Permeate is extracted via a tube running through the centre of each sheet, ensuring a very short permeate flow path. The company has also patented the Mechanical Cleaning Process (MCP), developed jointly with Darmstadt Technical University and Osnabrueck University of Applied Sciences, based on plastic granules which are added to the sludge which provide supplementary mechanical cleaning enabling higher fluxes and reduced chemical cleaning. The company launched the *XL* version of the *BIO-CEL*® in May 2014, which provides a total membrane area of 1,900 m², considerably higher than any other commercially-available single FS membrane cassette at that time.

The company had 11 industrial effluent references to 2013, with a total installed capacity of 3.5 MLD, three of which are fitted with the MCP. Applications include food and beverage, textile and laundry effluent.

4.1.17 newterra

newterra, an Ontario-based company, acquired the German membrane company Weise in 2012, whose well-established *MicroClear*® MBR membrane module was originally introduced in 1997. The modules, which are relatively small, are characterised by a permeate manifold (or "filtrate collector") which is fitted along the length of one side, providing a very short permeate flow path. Membrane cassettes (or filter housings) are offered in a wide range of configurations. The *MB* range of filter housings is based on an 8 m²-area 500 mm-high *MCXL2* module which can be stacked 2-5 deep, 1-5 long and 1-2 wide in frames to form the complete unit (Table 4-17). The

20 different filter housings offered for the *MB* range provide membrane areas ranging from 80 to 1,000 m². There are additionally the smaller *MA* module range and the so-called "high performance" *MX* range, the largest of which has a membrane area of 1,600 m².

Table 4-16 MICRODYN-NADIR *BIO-CEL®* membrane and module specifications

Material Pore size, µm:	PES 0.04 (150 kDa)				
Model:	***BC10F-C10***	***BC50F-C25***	***BC100F-C25***	***BC400F-C100***	***BC XL***
Height, mm:	1610	1563	1563	2763	2650
Length*, mm:	154,5	694	1270	1298	2800
Width, mm:	610	702	702	1152	2100
Membrane area, m^2:	10	50	100	400	1900
Packing density, m^2/m^2:	106	102	113	268	323
Max membr. air scour rate, $Nm^3/(m^2.h)$:		0.60		0.35	0.35
Flux, LMH: Recommended operating TMP, mbar	10-40 30-400				

* Length of base frame without pipework

Table 4-17 newterra *MicroClear® MB* module and filter housing range

Material Clean water permeability, LMH/bar: Pore size, µm:	PES or PVDF >200 0.04 or 0.3					
Model:	***MB2-1***	***MB2-5***	***MB3-1***	***MB3-5***	***MB5-1***	***MB5-5***
Height, mm:	1350	2350	1850	1850	2950	2950
Width, mm:	700	700	700	2380	700	2380
Thickness or breadth, mm:	1300	2380	1300	1300	1300	1300
Membrane area, m^2:	80	400	120	600	200	1000
No. cassettes per frame:	10	50	15	75	25	125
Packing density, m^2/m^2:	88	240	132	194	220	323
Membrane air scour rate, $Nm^3/(m^2.h)$:	0.31-0.63		0.21-0.42		0.32-0.63	
Max recommended flux, LMH:	25-30; 50 for short-term peak flows					
Recommended TMP range, mbar:	100-250 filtration, 100 backflush					

Frames available for five configurations within each range

4.1.18 Pure Envitech

Pure Envitech Co., Ltd. is a Korean membrane company established in 1996, providing both pure and wastewater products. Their MBR products comprise both the conventional "long panel" type (*ENVIS*) and the "cinder block" type (*SBM*, or "submerged block membrane") design. The latter comprises 15 rectangular panels, smaller than the traditional long panels and horizontally oriented to form a block 195 mm high and around 130 mm wide (Table 4-18). Stacks of up to 10 blocks can be produced by sliding them into a retaining metal frame, making the overall stack just over 3 m tall. The "cinder block" arrangement apparently increases the packing density by over 30% with respect to the projected module floor area, though the 10 block-high module is slightly taller than the *ENVIS* double-deck.

4.1.19 QUA

QUA Group, based in the United States and with offices in India, China and Italy, provides membrane technologies which include FS and HF synthetic membranes, a ceramic membrane and an electrodeionisation technology. The PVDF-based product forms the basis of the company's *EnviQ*™ iMBR FS technology (Table 4-19), which is based on a rack modular design using 10 m² membrane blocks (or cartridges) stacked four-deep within the frame.

4.1.20 SINAP

Shanghai SINAP Membrane Tech Co., Ltd. began its research and development into UF membranes in 1980, establishing the membrane technology R&D centre in 1992 as the

"Shanghai Institute of Applied Physics". The company offers four panel sizes based on a 0.1 μm PVDF membrane, with the panel membrane area ranging from 0.1 up to 1.5 m^2 (Table 4-20). The two smallest sizes (340 x 470 mm and 220 x 320 mm) are primarily for testing purposes. The technology has been applied across a number of industrial sectors, including to oil-bearing effluents (relating to steel foundry effluent), laundry, pharmaceutical and landfill leachate wastewaters. The company has more than 90 municipal and 130 industrial installations providing a flow capacity of almost 20 and >50 MLD respectively.

Table 4-18 Pure Envitech membrane, panel and block specifications

Material: Pore size, μm:	Chlorinated PVC 0.4					
Model	***ENVIS***			***SBM***		
	Panel	Module, s	Module, d	Element	Block	*SP* stack
Height, mm:	500-1500	1010-1760	2720	195	195	3140
Width, mm:	560	680-685	720	709	709	800
Thickness or breadth, mm:	-	295-1795	1898	7	127	3370
Membrane area, m^2:	0.5-1.5	up to 150	-	0.4	3.9	780
Number of panels/blocks[a]:	-	up to 100	-	-	-	200
Packing density, m^2/m^2:		115	220		43	289
Membrane air scour rate, $Nm^3/(m^2.h)$:	-	0.48-0.72		-		
TMP max, mbar:	790					
Recommended flux range, LMH:	13-21			17-29		

[a]panels per module, *ENVIS*; elements per block, blocks per stack, *SBM*.

Table 4-19 The QUA *EnviQ*™ module specifications

Material Pore size, μm:	PVDF 0.04		
Model:	***E8C***	***E16C***	***E32C***
Length, mm:	660	990	1262
Width, mm:	520	680	990
Height, mm:	2315		
Membrane area, m^2:	80	160	320
Packing density, m^2/m^2:	233	238	256
Recommended flux range, LMH:	10-30		
Membrane air scour rate, $Nm^3/(m^2.h)$:	0.3-0.8		
Recommended max TMP, mbar	500		

Table 4-20 SINAP membrane and module specifications

Material Clean water permeability, LMH/bar: Pore size, μm:	PVDF 515 0.1			
Model:	***SINAP 80***		***SINAP 150***	
	Panel	Module[a]	Panel	Module[b]
Length or height, mm:	1000	1850	1750	2600
Width, mm:	490	720-755	490	720-755
Thickness or breadth, mm:	7	965-3100	7	965-3100
Membrane area, m^2:	0.8	40-160	1.5	75-300
No. panels/cassette:	-	50-200	-	50-200
Packing density, m^2/m^2:	-	58-68**	-	108-128
Membrane air scour rate, $Nm^3/(m^2.h)$:	0.48-0.9			
Flux, LMH:	16-25			

[a]single-deck
[b]includes footprint incurred by permeate manifold tube

(a) (b)

Figure 4-6 SINAP module (a) permeate collectors and manifold, (b) modules

4.1.21 Supratec

Supratec Filtration GmbH & Co. KG from Germany provides a module with possibly the highest volume packing density of all the FS modules (139 m^2/m^3), and consequently the lowest air scour rate (0.12-0.25 $Nm^3/(m^2.h)$). The membranes are provided as both PES and PVDF and are backflushable in 560 m^2 membrane modules 2,500 mm high by 2,300 mm long and 700 mm breadth. The recommended flux and TMP ranges are 10-32 LMH and 20-350 mbar.

4.1.22 Toray

Toray Industries, Inc., a Japanese membrane manufacturer, has been supplying RO and NF membranes since 1967, subsequently introducing its pressurised and then submerged UF and MF HF range for filtering freshwater and other low-solids feeds. The FS MBR membrane panel (*TSP50100*, Fig. 4-7a) and module (the *MEMBRAY*®) was launched in 2004. The product employs a 0.08 μm PVDF membrane which is reinforced with a PET felt and mounted on a 7.5 mm-thick ABS supporting plate. Two panels are offered (Table 4-21), one providing 0.9 m^2 and used for the *TMR090* module (Fig. 4-7b) and a longer one of 1.4 m^2 for the *TMR140* module (Fig. 4-7c). The shorter panel finds applications when constraints are imposed on the module height (such as on board ships), the total module height being ~1.5m.

The panels form single- and double-width modules, and the longer-panel module can additionally be stacked to reduce the footprint. Panels are assembled in a stainless steel frame to form modules ranging from 45 m^2 total membrane area (50 elements, *TMR090-050S* module, Fig. 4.7b) to 140 m^2 (100 elements, *TMR140-100S* module). The modules can then either be doubled in width (*TMR140-200W* module) or stacked (*TMR140-200D* module) to form larger modules. The *TMR140-100S* module has dimensions of 1,620 mm long, 810 mm wide and 2,100 mm high. A design flux of 33 LMH is assumed (though the quoted range is between 8.3 and 62.6 LMH for peak operation), along with a maximum TMP of 0.2 bar. The recommended aeration rate is 0.56 $Nm^3/(m^2.h)$ for the single-deck unit, and pre-screening to 3 mm is stipulated.

Since 2003 over 500 MBR plants worldwide fitted with Toray membranes have been installed or are under construction, giving a total installed capacity of over 550 MLD. Around 27% of these plants are employed for industrial effluent treatment, providing >100 MLD capacity. They are thus the second-largest FS MBR provider, and a number of US wastewater process technology suppliers offer MBR systems based on the Toray membrane.

Table 4-21 Toray *MEMBRAY*® membrane panel and module specifications

Material	PVDF				
Clean water permeability, LMH/bar:					
Pore size, μm:	0.08				
Model:	***TMR090***		***TMR140***		
	Panel	Module, s	Panel	Module, s	Module, d
Length or height, mm:	1059	1470	1608	2100	4160
Width, mm:	515	730	515	810	810-840
Thickness or breadth, mm:	13.5*	990-1720	13.5*	950-3260	1620-3260
Membrane area, m²:	0.9	45-90	1.4	70-280	280-560
No. panels per cassette:	-	50-100	-	50-200	200-400
Packing density, m²/m²:	-	62	-	91-106	204-213
Membrane air scour rate, Nm³/(m².h):	0.67		0.56		
Flux, LMH:	6-32				
Recommended max TMP, mbar	200				

*including panel separation

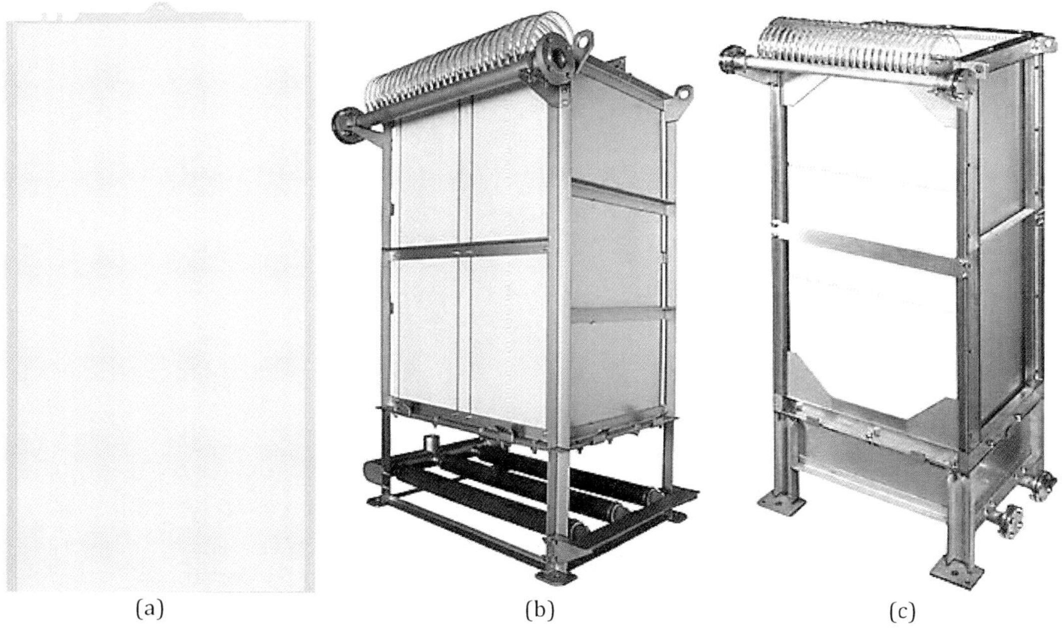

(a) (b) (c)

Figure 4-7 The Toray MBR: (a) *TSP50100* panel, (b) *MEMBRAY*® module (*TMR090*), and (c) *TMR140* module

4.1.23 FS product summary

The above listing is in no way comprehensive, and it is especially challenging to elucidate the commercially available products in China in particular (Section 4.4.1). There are undoubtedly more MBR membrane suppliers operating in China than is evident from open sources, but it is unclear as to how many of these provide original products. There are also examples of FS sidestream modules being used as MBR technologies (such as the ROCHEM technology, Section 4.3.1.5), but these are fairly unusual and their application appears to be largely limited to ships.

It is currently unclear as to whether the FS ceramic membranes, introduced since around 2008, will significantly penetrate the MBR market. Whilst ceramic membranes remain relatively expensive, and also appear to provide the lowest volumetric packing density (Fig. 4-8), the capital costs appear to have decreased significantly in recent years and their robustness to aggressive oxidative chemicals and elevated temperatures clearly make them attractive for some niche industrial effluent treatment applications. This being the case, it may be expected for other suppliers to emerge offering ceramic products. The composite membranes and the "cinder block" design, on the other hand, appear to provide the highest packing densities of the FS configurations. Low packing densities may be countered by higher new net fluxes, however. Packing density becomes a significant parameter when spatial restrictions exist and when existing infrastructure, and specifically secondary clarifiers, have to be considered (Section 4.5).

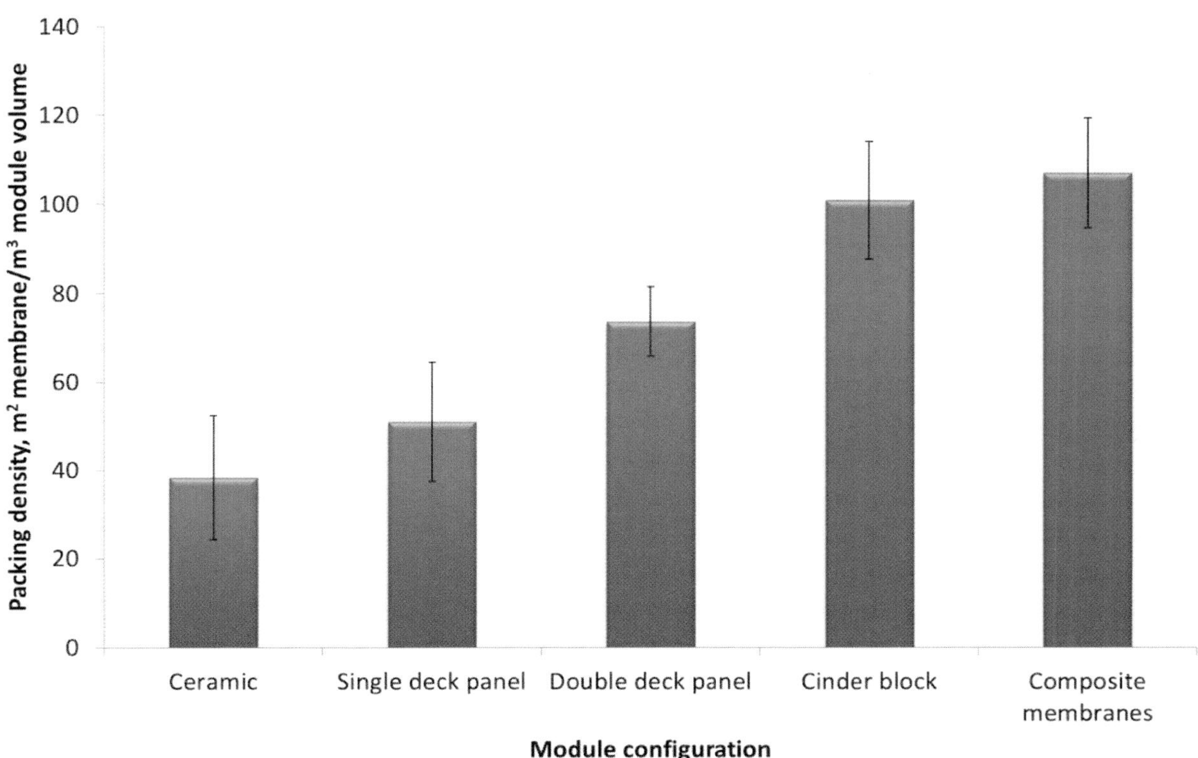

Figure 4-8 Summary of volumetric packing density, FS MBR membrane technologies

4.1.24 FS technology suppliers

There are a significant number of companies worldwide who provide MBR technologies based on a specific FS membrane product, or range of products, normally provided under license by some of the major membrane suppliers. The license agreement may be regional, application-specific, or both.

The largest of the MBR technology suppliers operating in the US and UK is Ovivo, formerly Enviroquip (in the US) and Eimco Water Technologies (in the UK) until acquired by GLV in 2006. Ovivo offer MBR systems based on Kubota and MICRODYN-NADIR membrane products. As an established supplier of Kubota membrane-based MBR technologies, Ovivo possibly has more references than any other specialist MBR technology supplier. In North America alone the company has 15 industrial and 174 municipal MBR references, alongside 31 membrane thickening plants based on the same FS technology. The company has also installed 15 industrial and 121 municipal plants and 15 thickening plants overseas. In total the company thus has 371 references, as at June 2014.

The US company Kruger has had an exclusive distribution agreement for the Toray FS MBR membranes in the US since 2007, offering the *NEOSEP™* process. A number of other MBR technologies in North America (for example the Smith & Loveless *Titan Membox*, ADI Systems Inc. and the Wigen MBR options) appear to be based on leading FS MBR membrane products. ADI Systems Inc., based in Canada, offer both aerobic and anaerobic MBR immersed technologies, with 21 aerobic iMBR (*ADI-MBR*) references and four of the anaerobic system (*ADI-AnMBR*).

Outside of North America there many tie-ins and joint ventures globally, including a significant commercial development in 2012 in the formation of LG-Hitachi Water Solutions, LG being the Korean giant and Hitachi already being an established provider of containerised MBR plants. There are a large number of companies providing specialist containerised technologies based on FS membranes, including Condor (UK), Busse (Germany), Hydroflux and OzMBR (Australia) and MİTTEM (Turkey). Specialist systems for ships are provided by Wartsila and Evac Oy, both based in Finland, and RWO (a Veolia Water Solutions company) in Germany, amongst others.

4.2 *Immersed hollow fibre (HF) products*

4.2.1 Asahi Kasei

Asahi Kasei Chemicals Corporation is active in both the municipal and industrial sectors in East/South-East Asia. The company produces the *Microza*® PVDF HF membrane for fresh and wastewater treatment applications, with a number of large plants installed for both applications. This 1.3 mm filament diameter membrane is widely established for freshwater applications, particularly where slightly increased resistance to the oxidative cleaning chemicals is required, and also forms the basis of the *MUNC-620A* MBR module. The module (Table 4-22) has been employed in China for petrochemical effluent treatment since 2006 (the Hainan Dao and Daya Bay projects), the Hainan project being possibly the first large (>10 MLD) MBR plant in China for this duty. The skid comprises 24 discrete cylindrical membrane bundles or elements (Fig. 4-9).

Table 4-22 Asahi Kasei *Microza*® *MUNC-620A* membrane module and skid specifications

Material:	PVDF	
Clean water permeability, LMH/bar:	-	
Filament diameter, mm:	1.3	
Pore size, μm:	0.1	
	Element	Skid
Length or height, mm:	2163	2900
Width or diameter, mm:	167	1400
Breadth, mm:	-	920
Membrane area, m²:	25	600
No. module per skid:	-	24
Packing density, m²/m³:	896.41	466
Membrane air scour rate, Nm³/(m².h):	0.2-0.28	
Recommended max TMP, mbar	800	

Figure 4-9 Asahi Kasei module

4.2.2 Econity

Econity Co., Ltd., formerly Korean Membrane Separation until renamed in 2010, is possibly Korea's most established membrane supplier having formed in 1998. The company provides both pressurised and immersed membranes. Their MBR membrane technology is one of the few based on "blocks", which contain 13 small ~38 mm-high "sub-units" of ~1.3 m² membrane area. The 0.65 mm-diameter membrane filaments are of hydrophilicised HDPE, one of the more unusual membrane materials produced by extruding the polymer to generate elliptical pores of 0.4 μm, as well as a 1.1 mm-diameter PVDF HF with 0.1 μm pores. The blocks are light enough to be installed manually in stacks up to seven deep within a metal cassette (Table 4-23).

The company's MBR technology based on the *CF* module is marketed as the *KSMBR*. As of April 2014, there were more than 1,700 municipal and 260 industrial installations based on the Econity technology, the respective flow capacities being almost 650 and ~67 MLD. The corresponding total installed membrane area is close to 2.5m m². References include some substantial installations at industrial parks, for example the 8 MLD plant at Myeonghak and the 25 MLD plant at Dalsung (Section 5.8.1).

Table 4-23 Econity membrane, cartridge and cassette specifications

Material:	HDPE, PVDF		
Clean water permeability, LMH/bar:	-		
Filament diameter, mm:	0.65(HDPE), 1.1(PVDF)		
Pore size, μm:	0.4(HDPE), 0.1(PVDF)		
Model:	***CF series***		
	sub-unit	cartridge	cassette
Height, mm:	-	396	1520-2770
Width, mm:	-	320	784-2264
Length or breadth, mm:	-	536	605-1300
Membrane area, m^2:	1.3	17	101-1210
No. module per skid:	-	-	6*-72
Packing density, m^2/m^2:	-	99	232*
Membrane air scour rate, $Nm^3/(m^2.h)$:	0.15-0.21		
Flux, LMH:	12.5-20		
Recommended TMP range, mbar	50-600		

**CF-23D* cassette, three-deep stack

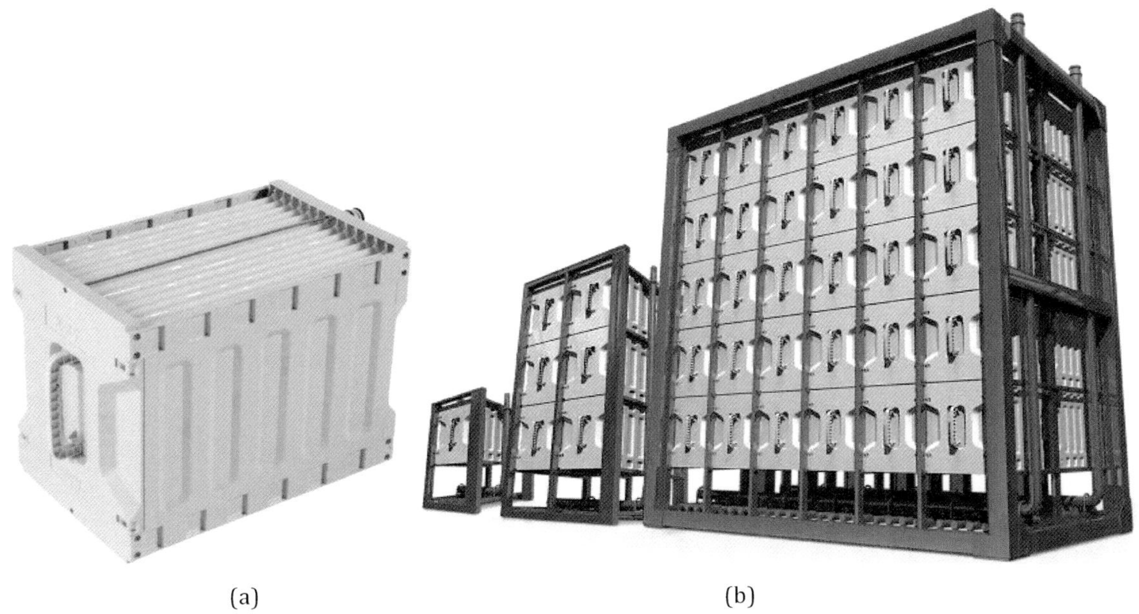

(a) (b)

Figure 4-10 Econity (a) cartridge and (b) stack (or cassette/frame)

4.2.3 Evoqua

Evoqua Water Technologies, LLC are the latest owners of the *MEMCOR*® membrane product range, previously owned by Veolia, US Filter and Siemens. The company provides both the pressurised and immersed modules for freshwater and wastewater applications, and is well established in the potable water area. The immersed *B40N* membrane module (Table 4-24) used for the MBR technology (the *MemPulse™*) is based on a 0.04 μm, 1.3 mm-diameter PVDF membrane which is intermittently aerated via an "air box" fitted at the base. The company has more than 100 MBR reference installations globally, primarily in municipal wastewater treatment.

Table 4-24 Evoqua *MemPulse™ B40N* membrane module and skid specifications

Material	PVDF	
Filament diameter, mm:	1.3	
Pore size, μm:	0.04	
Model:	***B40N***	
	Module	Rack
Height, mm:	1600	2261
Width, mm:	203	261
Length or breadth, mm:	203	3862
Membrane area, m^2:	38	608
No. modules per cassette:	-	16
Packing density, m^2/m^2:	922	603

4.2.4 GE

GE Water & Process Technologies (formerly Zenon) introduced the *ZW500D™* module (Table 4-25) in 2002, having continuously made improvements to the design of the *ZW500C™* (2001), and prior to that the *ZW500A™* (1997). These both followed commercialisation of the original "Zeeweed" module, the *ZW145™*, in 1993. This was the first HF MBR module to be introduced based on a "braided core" whereby the fibre is reinforced internally, and this structural element remains integral to the current *ZW500D™* product design. The GE technology remains the most widely implemented globally. The estimated global capacity of MBRs based on GE membrane technology is greater than 4,800 MLD, with about 15% of this being dedicated to industrial effluent treatment. GE's most recent offering is *LEAPmbr™*, which employs the latest *ZW500D™* membrane module. This technology incorporates a novel aerator design for improved overall energy consumption and reduced complexity of the aeration ancillary equipment, along with several other process design optimisations which provide an increased packing density and further operational efficiencies.

Table 4-25 GE *ZW500D™* membrane module and cassette specifications

Material	PVDF	
Filament diameter, mm:	1.9	
Pore size, μm:	0.04	
Model:	***ZW500D™***	
	Module	Cassette
Height, mm:	2198	2561
Width, mm:	844	1745
Length or breadth, mm:	49	2112
Membrane area, m^2:	34.4	1651
No. modules per cassette:	-	48
Packing density, m^2/m^2:	831	448
Membrane air scour rate, $Nm^3/(m^2.h)$:	0.1–0.2	
Flux, LMH:	-	
Max TMP, mbar	550	

4.2.5 Hinada

Hinada Water Treatment Technology Co., Ltd. of Luogang, Guangzhou, offer submerged PVDF membrane modules based on a 0.1 μm PVDF fibre of unspecified diameter. Module products are offered in three sizes: the *1015* and *1010* modules both have dimensions of 571 x 45 x 815 (h) mm and are respectively 15 m^2 and 10 m^2 in membrane area, whereas the *1520* is taller at 1,535 mm. The *1520* is marketed as a "Submerged Anaerobic" module, with the same target flux range of 10-20 LMH as the other modules.

4.2.6 Hyflux

The Singaporean-based company Hyflux Ltd originally supplied the FS product *Petaflex* but now appear to offer an HF configuration *Porocep®* (Table 4-26) based on HDPE, one of the more unusual MBR membrane materials. The *Porocep*® system comprises ten 8.5 m^2 elements fitted into a 1 m-tall module (or "box"), with the *POR* modules then forming a single- or double-deck skid.

Table 4-26 Hyflux *Porosep*® membrane module and skid specifications

Material: Filament diameter, mm: Pore size, μm:	HDPE 0.4 0.1		
Model:		***POR 101 -510***	***POR 102-1020***
	Module*	Single-deck	Double-deck
Height, mm:	1000	1744	2748
Width, mm:	500	1050	1050
Length or breadth, mm:	345	1163	1163
Membrane area, m^2:	85	510	1020
No. module per cassette:	-	6	12
Packing density, m^2/m^2:	493	279	354
Membrane air scour rate, Nm^3/(m^2.h):	0.1 (double-deck) - 0.2 (single-deck)		
Net flux, LMH:	10-20 (with 10-30% relaxation)		
Recommended operating TMP, mbar	100-500		

*module comprises 10 x 8.5m^2 elements

4.2.7 Koch Membrane Systems

Koch Membrane Systems Inc. is an established membrane filtration technology supplier – dating back to the early 1960s – prior to acquiring the *PURON*® technology in 2004. The technology is unusual in that the aerator is integrated with the membrane module, the latter comprising a series of 3.47-4.56 m^2 cylindrical fibre bundles (Table 4-27) according to the bundle length. Nine such bundles form a "fibre row", and between 8 and 44 such rows form complete modules of between 250 and 1,800 m^2 membrane area. The Koch Membrane Systems *PURON*® product is also differentiated by having a single bottom header with the fibres individually sealed at the top end. This is to allow the sludge solids to escape from the top of the module without being impeded by the header. The HF membrane has a braided core to provide mechanical strength and was originally of PES until a PVDF *PSH* product was introduced in 2009.

Table 4-27 The Koch Membrane Systems *PURON*® membrane and module specifications

Material Filament diameter, mm: Pore size, μm:	PVDF 2.6 0.03		
Model:	Bundle	***PSH 31-41***	***PSH 250-1800***
Height, mm:	1821-2319	1821-2319	2384 -2530
Width, mm:	92	828	893-1755
Thickness or length, mm:	92	92	906-2244
Membrane area, m^2:	3.47-4.56	31-41	250-1800
No. bundles/rows per module:	-	9	8-44
Packing density, m^2/m^2:	410-542		309-457
Membrane air scour rate, Nm^3/(m^2.h):			
Flux, LMH:			
Max TMP, mbar	600 forward filtration & backflush		

4.2.8 Kolon

The Kolon Industries, Inc. PVDF HF membrane is used for both fresh and wastewater treatment. It is strengthened by a braided core, and is mounted in modules which are configured both horizontally and vertically (Table 4-28). A number of industrial complexes in Korea and China have MBR effluent treatment plants based on the Kolon module (called *Cleanfil*®); configured as an MBR it is referred to as the *KIMAS* (Kolon Immersed Membrane Advanced System) MBR. The MBR cassette can contain 10-30 horizontal modules or 20/40 vertical modules.

Table 4-28 Kolon *Cleanfil*® module specifications

Material	PVDF	
Filament diameter, mm:	2	
Pore size, μm:	0.1	
Model:	***S20H***	***S30V***
	horiz	vert
Height, mm:	626	2367
Width, mm:	1185	1060
Length or breadth, mm:	105	44
Membrane area, m^2:	20	30
No. modules per cassette:	10-30	20
Packing density, m^2/m^2:	-[a]	-[a]

[a] not specified

4.2.9 Litree

Litree Purifying Technology Company Ltd has two facilities in Hainan and Jiangsu province, and is possibly the largest and longest established (1992) of the Chinese HF UF membrane manufacturers with a very significant share of the UF market for municipal drinking water and industrial pure water supply in mainland China. Their latest product (the *GEMINI*), launched in 2013, is a reinforced (i.e. braided) HF UF PVDF membrane. The module is offered in three sizes: 14, 22 and 31 m^2 membrane area (Table 4-29).

Table 4-29 Litree *GEMINI* membrane, module and cassette specifications

Material	PVDF/PVC (braided)				
Filament diameter, mm:	2				
Pore size, μm:	0.02				
Model:	Module			Cassette	
	2000-RF	***1500-RF***	***950-RF***	***2000x52***	***2000x26***
Height, mm:	2122	1622	1082	2520	2710
Width, mm:	721	721	721	1620	805
Thickness or length, mm:	70	70	70	2520	2710
Membrane area, m^2:	31	22	14	1612	806
No. modules per cassette:	-	-	-	52	26
Packing density, m^2/m^2:	614	436	277	356	498
Flux, LMH:	15-60				
Max recommended TMP, mbar:	600				

4.2.10 Hangzhou Microna Membrane Technology Co. Ltd.

Hangzhou Microna Membrane Technology Co. Ltd has produced PP, PVDF and M-PA (modified polyamide) hollow fibre membrane products (Table 4-30) for 10 years. The company collaborates with Zhejiang University, and has a 4,000 m^2 production facility capable of supplying 200m m^2 membrane per year. Its products are used in water/wastewater treatment, the food and biotechnology industrial sectors and for the degassing of pure water.

4.2.11 Memstar/United Envirotech

Memstar Pte Ltd, based in Singapore, was established only in 2005. It became the fully owned subsidiary of Memstar Technology Ltd in 2007 when the company was listed on the Singapore Stock Exchange, and undertakes the entire membrane production business. Memstar Pte Ltd was acquired by United Envirotech Ltd, an MBR engineering specialist and fellow Singaporean company who are also highly active in the Chinese market, in April 2014. The company has achieved significant penetration of the Chinese market since its foundation, aided by two manufacturing facilities in China (Guangzhou and Mianyang). Memstar installations provide over 900 MLD of capacity: 467 MLD from 15 municipal MBRs and 481 MLD from 51 industrial MBRs. The *SMM* modules (Fig 4-11), based on a 1.3 mm-diameter PVDF HF material (Table 4-

31), provide one of the highest module packing densities of all the commercially-available MBR membrane products.

Table 4-30 Hangzhou Microna membrane and module specifications

Material:	PP	PVDF, reinforced	M-PA*, braided
Pore size, μm:	0.04	0.01	0.008
Height, mm:	850, 1050, 1600	1000, 1500, 1800	
Width, mm:	525, 620, 620	620	
Length or breadth, mm:	25, 32, 32	32	
Membrane area, m²:	15, 20, 30, 40	10, 15, 20, 25	
Max. no. modules per cassette:	30	25	20
Packing density, m²/m²:	230	228	225
Membrane air scour rate, Nm³/(m².h):	0.15	0.5	0.6
Flux, LMH:	5-12	10-30	10-25
Max operating/backflush TMP, mbar	300	300	600

*modified polyamide

Table 4-31 Memstar/United Envirotech membrane and module specifications

Material	PVDF					
Filament diameter, mm:	1.3					
Pore size, μm:	0.1					
Model:	***SMM-1010***	***SMM-1013***	***SMM-1520***	***SMM-1525***	***SMM-2027***	***1520 cass.***
Height, mm:	815	815	1535	1535	2040	1531
Width, mm:	571	571	571	571	571	575
Thickness or length, mm:	45	45	45	45	45	-
Membrane area, m²:	10	12.5	20	25	27	1760
Max. modules per cassette:	-	-	-	-	-	88
Packing density, m²/m²:	389	486	778	973	1051	-
Flux, LMH:	10-50					
Max recommend. TMP, mbar:	300					

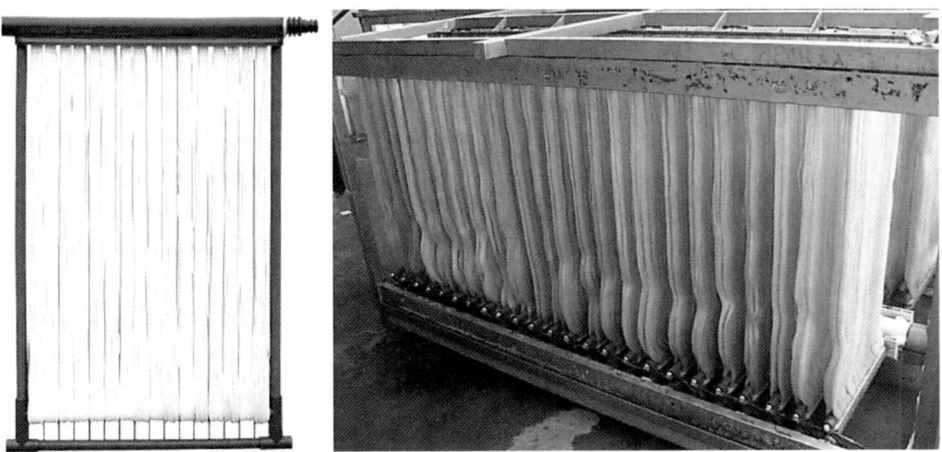

Figure 4-11 The Memstar module and cassette

4.2.12 Mitsubishi Rayon

Mitsubishi Rayon Co., Ltd provides the *STERAPORE™* range of MBR membranes. The company originally commercialised its horizontally-oriented *SUR* PE-based product in 1992, and is one of the top three largest MBR membrane suppliers worldwide in terms of global installed capacity. The company is possibly the largest in terms of installed capacity in East/South-East Asia, having introduced the *SADF* PVDF membrane in 2004 for the municipal market. There are now a number of large reference plants in China and Korea based on the *SADF* technology. In 2011 the modules were renamed and are now part of the *STERAPORE™ 5000* series. Since then the company has launched the *STERAPORE™ 5500* and *5600* series, which are upgraded follow-on models of the *5000* series (Table 4-32). There are two module sizes in the PE membrane range and four in the PVDF range.

Table 4-32 Mitsubishi Rayon modules and cassette specifications, *STERAPORE™ 5000* and *5500*

Material: Filament diameter, mm: Pore size, μm:	PE 0.54 0.4		PVDF 2.8 0.4		PVDF 2.8 0.05		PVDF 1.65 0.05	
Model*:	***2SL, 3SL***	***LP, LS***	***50E***	***50M***	***55E***	***55M***	***56E***	***56M***
	Mod.	Cass.	Mod.	Cass.	Mod.	Cass.	Mod.	Cass.
Height, mm:	1035	1442	1015-2000	1890-2800	1015-2000	1890-2800	2000	2800
Width, mm:	446-524	536-614	620-1250	650-1530	620-1250	650-1530	1250	1530
Length or breadth, mm:	14	1538	30	440-3740	30	440-3740	30	940-3740
Membrane area, m^2:	1.5, 3	105-210	6, 15, 25	18-1500	6, 15, 25	18-1500	40	400-2400
No. module per cassette:	-	70	-	3-60	-	3-60	-	10-60
Packing density, m^2/m^2:	-	128-223	-	62-262	-	62-262	-	278-420
Flux, LMH:	10-15		20-30		20-30		20-30	

*model numbers have been truncated for simplicity

4.2.13 Mohua

Mohua Tech Co. Ltd of Jiangsu province in China provides membranes for various applications. Its immersed PVDF HF modules are provided configured both as an array of small cylindrical bundles (the *iMBR-280* module within the *iMBR-900* frame) and as a conventional rectangular "curtain" (the *iMBR-25*). These are both around 2 m in total height.

4.2.14 Motian

As part of the Shandong Zhaojin Motian Co., Ltd. conglomerate, Motian provides a range of membrane technologies. Both PVDF and PP HF MF membranes are offered as immersed *FPA* modules (Table 4-33), these being of the conventional rectangular "curtain" type.

Table 4-33 Motian membrane module specifications

Material:	PVDF	PP
Filament diameter, mm:	1.2	1.8
Pore size, μm:	0.2	
Model	***FPA-13***	***FPA-14***
Height, mm:	1260	1800
Width, mm:	568	568
Thickness, mm:	45	45
Membrane area, m^2:	14	12
Packing density, m^2/m^2:	548	469
Recommended flux range, LMH	6-30	7-17
Recommended TMP range, mbar	20-500	

4.2.15 Motimo

Tianjin Motimo Membrane Technology Co., Ltd. is one of the largest UF/MF membrane manufacturers in Asia, extending back 40 years, and has been listed in the Chinese Stock Exchange since 2012. The company provides both pressurised and immersed membrane technologies for municipal and industrial wastewater reuse, water supply and industrial process water. The company provided the modules for one of the earliest MBRs for petrochemical effluent treatment based on a Chinese membrane (a 10 MLD plant installed in Sichuan, 2006). Motimo also provided the modules for one of China's largest MBR projects (a 100 MLD plant installed in Zhuzhou, 2014). The company's *FP* (*Flat plat*) PVDF membrane (1 mm filament diameter and 0.2 μm pore size) is offered as modules 1.5 m in length and having membrane areas of 12.5, 20 and 25 m^2.

4.2.16 Philos

Philos Co., Ltd. is a Korean membrane company established in 2002. Their *Megaflux* MBR membrane product (Table 4.34) is based on a braided PVDF filament configured as an array of cylindrical bundles of around 75 mm diameter. The bundles, which appear to provide a membrane area of around 0.87 m^2 per m length, are offered at three different module heights (1.1, 1.7 and 2 m) in five different modules (the *Megaflux 10, 30, 50, 70* and *100*, named according to their claimed flow capacity in m^3/d). The company also offers a sidestream HF module (*i-MBR*) which appears to be fitted with 16 bundles around a central permeate collector. This module operates with a significantly higher SAD_m than the conventional immersed HF module, and sludge is recirculated through it at a fixed rate of 4 m^3/h.

Table 4-34 Philos modules specifications

Material:	PVDF	
Filament diameter, mm:	2.3	
Pore size, μm:	0.4	
Model:	***Megaflux****	***i-MBR 14***
Height, mm:	1700-2000	2152
Width, mm:	710	-
Length or diameter, mm:	1070-1560	437
Membrane area, m^2:	-	30
Packing density, m^2/m^2:	-	200
Membrane air scour rate, $Nm^3/(m^2.h)$:	-	0.8-1
Recommended flux range, LMH:	-	10-40
Recommended operating TMP, mbar	-	66-158

*claimed capacity 50-100 m^3/d

4.2.17 Sumitomo

Sumitomo Electric Fine Polymer, Inc. is a large Japanese engineering company. Their MBR product (*Poreflon™*, Table 4-35) is one of the very few based on PTFE. The submergible membrane module, prefixed *05B,* is offered in two different heights, and is supplied as a tall, slim module roughly square in cross section. In the US the product is supplied by Layne Christensen, the first plant having been installed in 2003. By the end of the decade there were more than 200 plants based on the technology, for both municipal and industrial effluent treatment.

Table 4-35 Sumitomo *Poreflon™* module and frame specifications

Material:	PTFE		
Filament diameter, mm:	2.3		
Pore size, μm:	0.2		
Model:	***-05B6***	***-05B10***	***Frame****
Height, mm:	1520	2410	3900
Width, mm:	154	154	2280
Length, mm:	164	164	840
Membrane area, m^2:	6	10	200
No. modules per cassette:	1 - 64	1 - 64	20
Packing density, m^2/m^2:	238	396	104
Membrane air scour rate, $Nm^3/(m^2.h)$:			
Recommended operating TMP, mbar:		900	
Recommended backwash TMP, mbar:		2000	

*Example for 20 modules: frames can be provided for 1-634 modules

4.2.18 Superstring

The Superstring Membrane Technology Co. Ltd, PRC, *Super UF* membrane is 1.25 mm in diameter and is PP-based (Table 4-36). It is offered in a wide range of pore sizes (0.002, 0.2 and 1 μm) and with four cassette membrane areas based on 10-40 modules per cassette. The membrane product was commercialised in 2007.

Table 4-36 Superstring module specifications

Material:	PP			
Filament diameter, mm:	1.25			
Pore size, μm:	0.02, 0.2, 1			
Model:	***SuperUF-10***	***SuperUF-20***	***SuperUF-30***	***SuperUF-40***
Height, mm:	1124			
Width, mm:	264	558	852	1146
Length, mm:	760			
Membrane area*, m^2:	20	40	60	80
No. modules per cassette:	10	20	30	40
Packing density, m^2/m^2:	100	94	93	92
Recommended operating TMP, mbar	100-200			

4.2.19 Zena

The Zena system is based on rows of cylindrical bundles, similar to the Koch Membrane Systems *PURON*® (Section 4.2.7) and Philos *Megaflux* designs (Section 4.2.16). In this case, the membrane is based on a 0.1 μm PP filament (Table 4-37), which at 0.26 mm in external diameter is perhaps the thinnest of the MBR HF membrane products and must be wetted out prior to use. The product (designated *P5*) is provided in module rows of 10 or 12 cylindrical bundles which are 1.01 m in length and offer 0.8 m^2 membrane area. These are then assembled into modules containing up to nine rows to provide a maximum of 86 m^2 membrane area.

Table 4-37 Zena *P5* module specifications

Material	PP		
Filament diameter, mm:	0.26 mm		
Pore size, μm:	0.1		
Clean water permeability, LMH/bar:	150 (at 1 bar)		
	Row	***Module***	***Module***
Height, mm:	822, 1010	1010	1010
Width, mm:	643, 560	320	590
Thickness or length, mm:	31	560	784
Membrane area, m^2:	9.5, 8	40	86
No. bundles/module rows per module:	12, 10	5	9
Packing density, m^2/m^2:	477	223	186

4.2.20 HF product summary

A review of the iHF membrane products reveals that they are predominantly configured as rectangular sheets, or "curtains". The dimension of these curtains and the means by which they are formed into a module varies to some extent between products, with some modules being very narrow (e.g. Evoqua and Sumitomo) and others being of a more standard width of 0.5 to 1 m. Two to three products are configured as discrete bundles within a skid, such as the Asahi Kasei product (Figure 4-9), and another three to four have smaller cylindrical bundles connected in rows which then form the complete module (Koch Membrane Systems, Philos and Zena). Only one product (Econity) is configured as a sub-500 mm high "block" within a frame, permitting manual stacking in the same fashion as for the FS "cinder block" design.

There appears to be no consistent pattern in the packing density either with HF configuration (bundles vs. curtain) or, as might otherwise be supposed, filament diameter. Volumetric packing density for all but two of the products ranges from 115 to 250 m^2 membrane per m^3 module volume, the average value being 174 m^2/m^3 with a standard deviation of 19% - somewhat higher than the highest packing density values calculated for the FS technologies (Figure 4-8).

4.3 Sidestream multi-tube (MT) and multi-channel (MC) modules

There exist a few established polymeric multi-tube (MT) membrane products which have been employed for advanced biological treatment of industrial effluent, based on pumped sidestream MBR technology (Fig. 2.3c), since the 1980s. There are also a growing number of ceramic MT and multi-channel (MC) products (Section 4.3.2), though it appears these are not yet widely established as sMBRs. These pumped sMBR membrane modules are generally of a standard diameter (100 or, more usually, 200 mm diameter) such that they are, in effect, interchangeable provided they are also of a standard length (normally 1 m). Since around the turn of the millennium the air-lift or air-scoured sidestream technology (Fig. 2.3d) has been introduced by a number of technology suppliers (Section 4.3.3).

4.3.1 Polymeric MT membranes

4.3.1.1 Berghof

Berghof Membrane Technology, part of the family-owned Berghof Group, is one of the leading manufacturers of MT membrane products produced in Germany. Berghof offers both polymeric membrane products and MBR membrane technologies based on these membranes (Section 4.3.3.2). The MT membranes (termed *HyperFlux* products) are between 5 and 12.7 mm in internal diameter in both PES and PVDF, with pore sizes between 5 kDa and 0.03 μm, and module lengths of 1, 3 and 4 m. The 5 and 8 mm diameter membranes are backflushable. Modules are mounted either horizontally or vertically in a skid, depending on the technology, with area packing densities ranging from 100 m^2 membrane per m^2 floor area for the conventional pumped, horizontally-aligned modules to 114 m^2 per m^2 for the vertically-oriented air-lift module. The company had over 700 reference installations worldwide by the end of 2013, in a wide range of applications including landfill leachate, food and beverage, refinery, and pharmaceutical industrial effluents as well as municipal wastewater.

4.3.1.2 MEMOS

The Lichtenstein company MEMOS Membranes Modules Systems GmbH offers a wide range of MT UF products targeted at MBR applications. The modules (*MEMTUBE-CROSS*) are offered in PVDF, PES and PS at pore sizes between 30 and 250 kDa, tube diameters between 6 and 24 mm and module diameters between 65 and 250 mm and lengths up to 4 m. The maximum module area provided for the MT modules is 53 m^2 for a 250 mm-diameter module. The company additionally produces wide-bore HF modules (the *MEMTUBE-SUB*) for submerged applications.

4.3.1.3 PCI Membranes

PCI Membranes is a subsidiary of the Xylem group of companies, a global acting water company offering a wide variety of water technology product brands. PCI Membranes, a company originally established in the late 1960s, offers MT UF products ranging from 6 to 12.5 mm in diameter which are available in PVDF polymers. Membranes based on other materials (such as PES, PS, PAN, CA and various thin film composites) are also offered for niche industrial separation applications. Xylem Inc. also offers an immersed technology through its 2011 agreement with GE.

4.3.1.4 Pentair

The US company Pentair acquired the Clean Process Technologies division of the Dutch-based Norit Group of companies in 2011, and consequently includes the *X-Flow* range of membrane products amongst the many brands the company owns. Its activities are based mainly in water management and treatment, and encompass a wide range of industrial sectors including power, mining, oil and gas, food and beverage and manufacturing. The *X-Flow* product line is based on membranes operating with in-to-out flow, and includes a capillary tube (CT) – referred to as HF by the company – and an MT product which originated from Stork Friesland in the 1990s. The total global installed capacity of Pentair membrane-based MBRs is in the region of 500 MLD.

The MT membrane supplied for sMBR duties is based on a 0.03 μm pore size PVDF membrane mechanically supported by a robust substrate and a dual-layer polyester backing material. Products are offered both for conventional pumped systems and the newer air-lift configuration. The 3 m-long, 200 mm-diameter membrane modules offered for the pumped configuration comprise the *Compact 27* (Fig. 4-12a) and the *Compact 33* which respectively provide 27 m^2 and 33 m^2 of area from an array of 8 mm and 5.2 mm diameter tubes. The 33 m^2 module is also considered suitable for the company's *Airlift™* system (an A-LsMBR), for which the *Compact 32V* (32 m^2 per foot of membrane module length) is also offered (Fig. 4-12b). This product is based on 3 mm-diameter tubes and is used to form the *Megablock* module, comprising an array of 3 m-long vertical modules which are relatively tightly packed to provide a high area packing density.

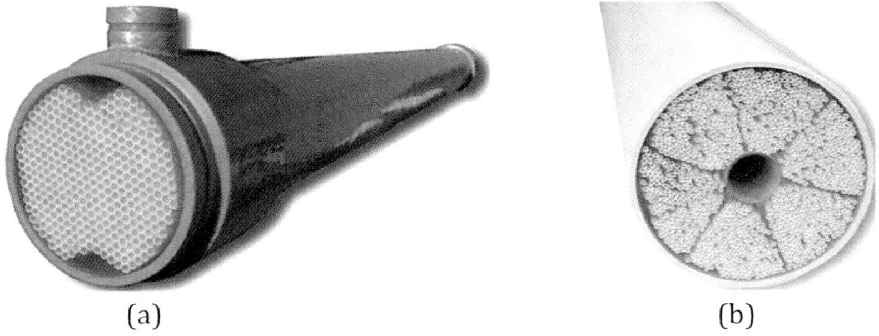

(a) (b)

Figure 4-12 The Pentair *X-Flow* modules (a) *Compact 27G* (GRP version of the *Compact 27*), and (b) *Compact 32V*, for the *Megablock*

The *X-Flow* MT membrane has been used by a number of process designers for proprietary MBR technologies, the pumped systems primarily dedicated to relatively small industrial effluent applications. The *Airlift™* configuration is more appropriate for relatively low organic concentrations and higher flows (<1 g/L COD and >250 m^3/hr), and this includes industrial and municipal effluents, whereas at high COD levels (>5 g/L) and low flows (<100 m^3/hr) the pumped system is recommended (Table 4-38). This leaves a range of flows and COD levels for which either system may be suited, but the overall COD loading rate over the entire range is normally between 100 and 1,000 mg/L COD/hr.

Table 4-38 Pentair *X-Flow* sMBR operational parameters

Parameter	*Pumped*	*Air-lift*
MLSS, g/L	12-30	8-12
TMP, bar	1-5	0.05-0.3
Flux, LMH	80-200	40-65
Permeability, LMH/bar	40-80	150-600
Footprint, m^3/hr capacity per m^2 projected area[1]	10.8[3]	7.5[2]
Footprint, m^2 membrane per m^2 projected area[1]	108	162
Specific energy demand for membrane filtration, $E_{L,m}$, kWh/m^3	1.5-3.5	0.25-0.5
Processing	Not backflushed	Backflushed

[1]based on a single skid, 1 m x 4 m x 4 m high. [2]based on 55 LMH maximum gross flux, 40-48 LMH net flux: a "staggered *Airlift™* skid" is 1.7 x 3.6 m footprint and contains 990 m^2. [3]based on 100 LMH.

4.3.1.5 ROCHEM

ROCHEM, part of the Ultura group, offers what appears to be a unique product (the *FM* module) comprising a small 0.68 m^2 FS cassette, with 3 mm spacing between the membrane sheets, of square cross section, fitted into a standard 200 mm diameter tube to form a module. Up to 10 cassettes can be fitted per module to provide the required conversion (up to 85%, with retentate recycling) with a maximum of 3 modules per skid, the mean flux attained being 30-40 LMH. The technology is employed in marine applications as part of the company's *Bio-FILT®* MBR system. The company has provided around 65 systems for ships, including cruise ships, yachts, research ships, navy vessels and support vessels, with capacities of up to 500 m^3/day.

4.3.2 Ceramic products

Ceramic MT/MC membranes are currently based on five different materials: aluminium oxide (as α-alumina, Al_2O_3), silicon dioxide (silica, SiO_2), titanium dioxide (titania, TiO_2), zirconium oxide (zirconia, ZrO_2), and silicon carbide (SiC). In principle, any of the identified commercial products (Table 4-39) can be used as a pumped sMBR technology. Individual suppliers are able to provide products over a wide range of nominal pore sizes, from ~1 kDa to ~1 μm, with different pore size ranges being attained using different materials.

Table 4-39 Ceramic membrane specifications

Product	Al_2O_3	TiO_2	SiO_2	ZrO_2	SiC	Geom[1]	Pore size	Spec area[2]
Ceramem (VWS)	x	x	x		x	S	0.05, 0.2 μm	0.22-0.46
Induceramic						C	450 Da-1.2μm	0.22-0.28
Likuid Nanotek	x	x		X		C	0.001-0.8μm	0.05-0.28
LiqTech					x	C, S	0.04-3μm	0.3-0.42
Tami	x	x				V	0.002-1.4μm	0.14-0.51
Atech	x	x		X		C	1k Da-1.2μm	0.05-0.28
Membralox (Pall)	x	x		X		C	0.02-1.4μm	0.24-0.36

[1]square (S), circular (C) or various (V) geometries. [2]specific membrane surface area per m length for a 25 mm-diameter tube. VWS Veolia Water Systems.

4.3.2.1 Grundfos BioBooster

One of the more unusual MBR technologies is that offered by BioBooster, an affiliate of the Danish firm Grundfos. The BioBooster membrane module is based on a stack (or "wagon") of 36 rotating 0.2 μm pore disc membranes, ~500 mm in diameter, which can be either ceramic or polymeric. Up to five of these are placed in a 6,300 mm-long tubular vessel ("Membrane Filtration Unit", MFU) to provide 34 m^2 area (and hence 0.19 m^2/disc). The MFUs are then assembled to form a skid (Fig. 4-13) 11,500 mm long by 2,480 mm high. For a rotation frequency of 80-180 RPM the power demanded is 3 kW. Thus, for fluxes above 40 LMH the *SED* for membrane rotation is 0.5-1 kWh/m^3 depending on the flux.

Figure 4-13 The Grundfos BioBooster stack

4.3.3 sMBR technology suppliers

MT technologies are offered by a number of different OEM suppliers, generally in three configurations:

a) Conventional pumped (horizontal membrane modules)
b) Low-energy pumped (horizontal membrane modules)
c) Air-lift or air scour (vertical membrane modules).

The conventional pumped configuration is employed for anaerobic systems (AnsMBRs) as well as for aerobic ones. It incurs the highest membrane separation energy demand of all the MBR technologies, but also the lowest footprint because of the high flux attained (in excess of 150 LMH for the most treatable effluents such as those from vegetable processing). The more recent low-energy pumped system is based on exactly the same configuration but employs a lower CFV and a commensurately lower operating flux. This decreases the flux but also reduces the energy demand. The air-lift configuration (A-LsMBR) permits lower-energy operation of the MT/MC module, whilst also demanding the pumping of both the air and the sludge to maintain operation of the membrane. All sMBR technologies allow energy management through the shutting down of some of the modules during periods of low flow and restarting when normal/high flow commences.

Since MT membranes are standardised, provided most commonly as 200 mm-diameter modules, they can be interchanged as desired. A number of different names are assigned to the proprietary technologies associated with the general configurations a)-c) above, those offered by four of the OEM sMBR suppliers being listed in Table 4-40 below. Of those suppliers listed, one offers all three configurations. All these suppliers tend to focus their activities on industrial effluent treatment and reuse applications and, in the case of the latter, all are usually able to provide both the sMBR and the downstream NF or RO plant for the complete reuse system.

Table 4-40 sMBR suppliers and technologies

Supplier	*Pumped*	*Low-energy pumped*	*Air-lift*
Aquabio	*AMBR™*	*AMBR LE™*	-
Berghof	*Bioflow*	*Biopulse*	*BioairDS*
Dynatec	*HiRate™*	-	*Dynalift*
WEHRLE	*HS (BIOMEMBRAT®)*	-	*LS*

4.3.3.1 Aquabio/Freudenberg

The UK company Aquabio Ltd, with more than 15 years experience of MBR installations, was acquired by the German filtration specialists Freudenberg Filtration Technologies in 2013. As with most of the OEM MBR technology suppliers, Aquabio's primary focus is on the industrial effluent treatment market, in particular food and beverage effluents. Other applications have included biofuels, pulp and paper, leachate, pharmaceutical and tannery effluents. The majority of their installations in food and drink wastewater treatment have involved additional RO treatment for water recovery.

The two different configurations – *AMBR™* and *AMBR LE™* (low-energy crossflow) – employ different operating conditions and thus have different attainable fluxes (Table 4-41). The company offers sMBR technology for both aerobic and anaerobic treatment. Aerobic treatment, combined with membrane separation to provide the sMBR technology, encompasses either slot or jet aerators for combined enhanced oxygen transfer and mixing. Pure oxygenation has also been used for some aerobic applications. Anaerobic applications combine the sMBR technology with the company's own *JETOX™* mixing technology and focuses on high-strength industrial wastewater (>10g/L COD).

Table 4-41 Aquabio/Freudenberg MBR technologies

MBR Type	*Normal MLSS range (g/L)*	*Sustainable flux LMH*	*SED_m, kWh/m³ permeate produced*	*Membrane orientation*
AMBR™	10 - 20 (air) 15 - 35 (pure oxygen)	80 – 250	1.8 - 3.5	Horizontal
AMBR LE™	10 - 20	40 – 120	0.4 - 1.5	Horizontal

A key feature of the *AMBR LE™* technology is the inclusion of variable speed recirculation pumps, which allow the system to revert to high crossflow operation during periods of high flows. The system otherwise operates at reduced CFV with backflushing to maintain a moderate flux and specific energy demand. Filtration is controlled by the sludge level in the bioreactor, and the filtration mode by the inlet flow or by the RO demand for water supply for reuse. The process is thus able to adapt to variable flows and loads and is thus suited to applications involving seasonal, diurnal or process-related variations to wastewater flows and concentrations, especially when unit power costs are significant. The process can be based on either standard 8" (200 mm) diameter modules or the more recently developed 4 m-long 10" (250 mm) modules which offer double the available membrane area and an increased back-pressure threshold (0.5 bar) for more vigorous backflushing.

4.3.3.2 Berghof

Berghof Membrane Technology offers three MBR technologies (Table 4-40). In the conventional pumped *Bioflow* system for high-concentration industrial wastewater applications a crossflow of 3–4 m/s is used to maintain the flux. The low-energy self-regulating *Biopulse* process, recommended for less complex and medium-strength wastewater, employs 5- or 8 mm-diameter backflushable PVDF *HyperFlux-LE* membrane modules and lower crossflows (1–2 m/s). The most recent *BioairDS* technology is designed primarily for municipal wastewater treatment, along with other less complex and less concentrated effluents. It employs vertical *HyperFlux-I8LE* or *HyperFlux-I5LE* modules with an air distributor (*Distair*) fitted at the top of the module to inject air simultaneously into each individual membrane tube at a fixed and precise rate, providing an optimum and even distribution of air and sludge.

4.3.3.3 Dynatec

The New Jersey-based company Dynatec Systems Inc. was established in 1979, its original focus being on physical membrane filtration of industrial effluents. In 1999, the company installed its first MBR, subsequently extending its operations to municipal as well as industrial effluents. The company uses MT membranes from a number of different suppliers for its pumped (*HiRate™*) and air-lift (*Dynalift*) technologies. Examples of industrial applications have included landfill leachate, including those containing hazardous substances, and food, automotive factory and metal processing plant wastewaters.

4.3.3.4 WEHRLE

Since 1982 WEHRLE Umwelt GmbH has engineered and built plants for the treatment of industrial effluents and complex wastewaters arising from the waste industry, including landfill leachate and digestate. WEHRLE also offers related services such as consultancy, plant operation, plant efficiency optimisation and upgrades to existing plants. In addition WEHRLE offers systems for biogas generation, sludge reduction and the recovery of heat energy and nutrients from effluent.

WEHRLE implemented crossflow sMBR technology (now termed the *HS* technology, as distinct from the *LS* low-energy pumped sMBR, Table 4-40) for landfill leachate treatment as early as 1990 at the Billigheim landfill leachate treatment plant. The company went on to develop its first air-lift MBR at Freiburg landfill in 1998, and 2004 saw the development of its first low-energy crossflow system using reduced crossflow velocity and back-flushing to minimise energy demand. Since then WEHRLE has built six additional low-energy crossflow systems and has over 300 reference installations, with a strong focus on the supply of turnkey crossflow MBRs.

4.3.4 Anaerobic sMBRs

The commercial history of the anaerobic sidestream MBR (An sMBR) process extends back to the Japanese Aqua-Rennaissance programme from throughout the mid-late 1980s. Pilot-scale tests (<10 m^3/d) were conducted on a number of different industrial and municipal wastewaters during this period and the early part of the 1990s. At around this time the South African company Wier En Vig introduced the *ADUF* (anaerobic digestion ultrafiltration) process based on Pall *Membralox* ceramic MC membranes. The process was also taken up by the company Bioscan in the 1990s (the *Biorek* process), primarily for piggery wastewater treatment in their native Denmark, and subsequently by the Veolia Water Systems subsidiary Biothane. Biothane's *Memthane* process was commercialised in May 2012, following the development of the technology in collaboration with Pentair. Whilst the *ADUF* and *Biorek* technologies no longer exist commercially, as of April 2014 there were 8 *Memthane* plants installed worldwide.

Anaerobic treatment is normally viable only at very high COD concentrations (15-250 g/L) when the recovered methane generated by anaerobic degradation of the organic matter provides a significant cost benefit (Section 2.3.4). The latter may then offset the increased costs associated with low flux operation (15-30 LMH), the longer residence times (necessitating a larger bioreactor), and the requirement for a sealed system.

4.4 Other products and technologies

4.4.1 Products from China

Whilst the listings provided in Sections 4.1-4.3 include a number of Chinese products, this should not be regarded as comprehensive. The Chinese market itself is very large, dynamic and complex.

A significant number of international MBR membrane product suppliers have agents or partners in China. According to a report published in 2013 (Marketsnresearch, 2013) these include Asahi Kasei, Evoqua, GE Water & Process Technologies, Huber, Hydranautics/Mitsubishi Rayon, Memstar, Kubota, Pentair, and Toray (via Bluestar). These companies (predominantly Japanese) nearly all have their bases in either Beijing or Shanghai. There are additionally a number of regional Chinese MBR membrane supplier companies, for which providing a comprehensive listing would be extremely challenging, which appear to provide mainly HF membranes. These include (city/province): BIC (Beijing), Creflux (Zhejiang), Dehong (Shanghai), Hina (Guangzhou), IWHR (Beijing), Keji (Shanghai), Feitian (Zhejiang), H-Filtration (Hangzhou), Jiamiao (Baoding), Jie Fu (Hangzhou), Kaihong (Zhejiang), Kaijie (Zhejiang), Qiangsheng (Zhangjiagang), Suntar (Xiamen) and Vina (Suzhou). Of these only BIC, IWHR, Suntar and Vina appear not to be configured as HF membrane modules.

Elucidating the precise capabilities of these regional Chinese membrane supplier companies is made challenging by inconsistencies and/or ambiguities in the information provided. Firstly, there is an issue of nomenclature. The names (or, possibly, just the spelling) of the products change – perhaps arising from issues of translation into English – and the companies themselves frequently change names, possibly associated with their (semi)privatisation if formerly government-owned. Secondly, the actual provenance of the product is rarely evident, such that the suppliers may not be offering original products. This difficulty is compounded by the lack of precise product specifications and images. Finally, it is unclear as to whether the products are actually configured (and used) as MBR modules or for conventional physical membrane filtration (as would be the case, for example, for polishing of secondary wastewater, Fig. 2.1b).

What is certainly the case, however, is that the size of the Chinese market is considerable. A recently-published survey (GWI, 2014) of the global low-pressure membrane market has indicated that the Chinese market is set to exceed that of North America in terms of the combined capacity of large UF/MF membrane plants (in this case those plants greater than 10

MLD in treatment capacity). Moreover, the same survey showed the combined capacity of large industrial process water treatment plants in China to be in excess of 1,500 MLD, whilst the Chinese wastewater treatment capacity in total was above 50,000 MLD. Both these figures exceed the corresponding combined totals reported for North America, Singapore, France, Australia, South Korea and India. China thus has a considerable number of large membrane installations associated with the country's substantial industrial activity.

4.4.2 Other emerging technologies

Other MBR technologies exist which are primarily versions of the three main configurations, with the membrane product provided under license (Section 4.1.24). The recent introduction of ceramic FS MBR membrane products may pave the way for further commercialisation of similar technologies. Other technological developments may be driven by joint ventures or acquisitions, of which there have been number since the mid-2000s (Table 4-42) with the *MEMCOR*® technology twice changing hands in the space of 10 years. Most of the leading MBR technology suppliers are distinguished either by being part of larger business groups or else are large enough themselves to develop their own products.

Table 4-42 MBR products, major acquisitions

Buyer/target	*Countries*	*Date product launched*	*Date company acquired*	*Date, first >10 MLD plant*
GE/Zenon	US/Canada	1993	2006	2002
Koch Membrane Systems/Puron	US/Germany	2001	2004	2011
Pentair/Norit	US/Netherlands	2002	2011	2012
Siemens/Veolia	Germany/France	2002	2004	2008
Evoqua/Siemens	US/Germany	2002	2014	2008
United Envirotech/Memstar	Singapore	2005	2013	2010
Sinomem/MICRODYN-NADIR	China/Germany	2005	2006*	-

*Joint Venture/50% acquisition

Given the predicted growth of the wastewater treatment/reuse market generally and that of the MBR market specifically, it seems likely that new technological innovations will arise. The combination of membrane separation with moving bed bioreactor (MBBR) technology has been demonstrated by various research centres for a number of years, culminating in the introduction by GE in 2014 of its *MACarrier* based on such a hybrid system. This technology is targeted at industrial effluents containing significant levels of recalcitrant organics which the carrier helps to retain in the bioreactor, thus extending the retention time for biotreatment. It seems likely that other similar hybrid technologies will be commercialised in the coming years.

4.5 Appraisal

A listing of the membrane products available at the time of publication (Table 4-43) suggests that there are around 63 suppliers of original MBR membrane products, two suppliers offering more than one membrane configuration, and with the HF outnumbering the FS suppliers. 60% of the suppliers are located in China/Taiwan, Korea, Singapore and Japan (Fig. 4-14a), and the HF "curtain" configuration appears to be the most favoured overall (Fig. 4-14b). Whilst the number of products derived from European countries appears to be significant, these are nearly all for niche or regional markets and do not represent a significant proportion of the market in terms of installed capacity. Almost all of the Chinese products are targeted at the significant regional market, with only a few having the global reach and impact of some of the leading Japanese and North American brands.

Table 4-43 Commercial MBR membrane products

Immersed (iMBR)		**Sidestream (sMBR)**
Flat sheet	Hollow fibre	Multi-tube/multi-channel, polymer
A3/*MaxFlow*^{DE}	Asahi Kasei - *Microza* ^{JP}	Berghof - *HyPerm-AE; HyperFlux*^{DE}
Alfa Laval - *Hollow Sheet*^{SE}	Creflux^{CN}	Pentair – *Compact*^{US}
Beijing IWHR - *Gyroreactor*^{CN}	Dehong^{CN}	MEMOS - *MEMCROSS*^{DE}
Benenv^{CN}	Econity - *KSMBR*^{KR}	Xylem/PCI Membranes^{US}
Brightwater/Anua - *Membright*^{IRL}/*PuraM*®	Evoqua - *MemPulse*^{US}	
Ceraflo^{SG}	Feitian^{CN}	Multi-tube/multi-channel, ceramic
Ecologix - *EcoPlate*^{TN}	GE - *ZeeWeed* ^{US}	Likuid Nanotek^{SP}
Huber - *VRM*^{DE}	H-Filtration - *MR*^{CN}	Veolia Water Systems – *Ceramem*^{FR}
ItN^{DE}	Hina^{CN}	Suntar^{CN}
KOReD, *Neofil*^{KR}	Hinada^{CN}	LiqTech^{DK}
Kubota - *ES/EK*^{JP}	Hyflux - *Porocep*^{SG}	
Kubota - *SP*^{JP}	Jiamiao^{CN}	Flat sheet
Lantian Peier^{CN}	Jie Fu^{CN}	ROCHEM - *Bio-FILT*^{US}
LiqTech^{DK}	KaiHong^{CN}	Novasep^{FR}
Martin - *siClaro*^{DE}	Keji^{CN}	
MegaVision^{CN}	Koch Membrane Systems - *PURON*^{US}	Hollow fibre
Meiden^{JP}	Kolon - *Cleanfil*^{KR}	Mann+Hummel – *B100*^{DE}
MICRODYN-NADIR – *BIO-CEL*^{DE}	Litree - *LH3*^{CN}	Polymem - *IMMEM*^{FR}
newterra – *MicroClear*^{CA}	MEMOS - *MEMSUB*^{DE}	
Pure Envitech Co., Ltd. – *ENVIS*^{KR}	United Envirotech/Memstar - *SMM*^{SG}	Flat disc
Pure Envitech Co., Ltd. – *SBM*^{KR}	Microna^{CN}	Grundfos - *BioBooster*^{DK}
QUA - *EnviQ*^{US}	Mitsubishi Rayon - *STERAPORE 5000*^{JP}	
SINAP^{CN}	Mohua^{CN}	
Supratec^{DE}	Motian^{CN}	
Toray - *MEMBRAY*^{JP}	Motimo - *FP AIV*^{CN}	
Vina^{CN}	Philos^{KR}	
	Porous Fibers S.L. - Micronet ^{SP}	
	Qiangsheng^{CN}	
	Sumitomo - *Poreflon* ^{JP}	
	Superstring MBR Tech. Co.Ltd - *SuperUF*^{CN}	
	Zena^{CZ}	

CA: Canada; CN: China; CZ: Czech Republic; DE: Germany; DK: Denmark; FR: France; IRL: Ireland; JP: Japan; KR: Korea; NIR: Northern Ireland; NL: Netherlands; SE: Sweden; SG: Singapore; SP: Spain; TN: Taiwan; US: United States

With regards to the membrane material, there remains a strong preference for the PVDF polymer for all three of the membrane configurations and for the HF in particular; PES is also offered for the FS and MT configurations (Fig. 4-15). PES does not appear to be offered as a hollow fibre possibly because of its mechanical integrity limitations, it being a "glassy" polymer. Whilst there are currently few ceramic membrane installations providing only a very small installed capacity, there are nonetheless seven suppliers of such membranes offering eight products in total.

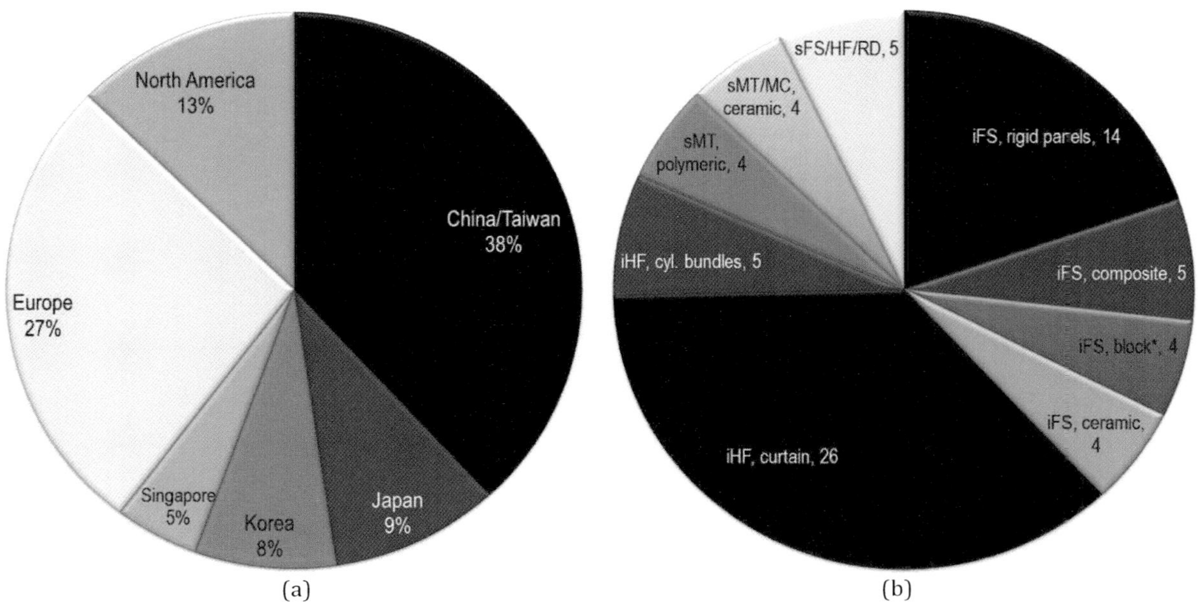

Figure 4-14 MBR membrane products (a) % by region, (b) number by configuration (*includes one composite membrane)

A key property of the membrane module is its footprint, or projected floor area. Footprint is a function of the area module packing density (m^2 membrane per m^2 floor area), the arrangement of the modules in the tank (for an immersed membrane) or skid (for a sidestream membrane), and the flux (Table 4-44). These three parameters then provide the equivalent of the overflow rate of a secondary clarifier, this being the flow rate per unit surface area in m/h.

Table 4-44 Surface overflow rates and corresponding footprint data

Parameter	Membrane				2ndary clarifier	
	sMT[1]	AlsMT[2]	FS	HF	min	max
Packing density, m^2/m^2	108	115-162	131	180	-	
Flux m/d	2.4	0.6	0.3	0.3	-	
Overflow m/d	259	97	39	54	16	28
Minimum % improvement	*826%*	*247%*	*40%*	*93%*		

[1]pumped sidestream MT MBR; [2]air-lift sidestream MT MBR

Table 4-44 shows example data for overflow rates for four different membrane technologies, using values for the overall packing density as calculated from the membrane specification data tabulated in Sections 4.1-4.3 and assuming appropriate values for the sustainable flux for each configuration. It is evident from this analysis that membranes offer a minimum 40% reduction in the footprint over that of a classical secondary clarifier basin, with a possible eight-fold decrease in footprint for a sidestream MBR operating at a flux of 100 LMH. This will clearly be instrumental wherever there is a constraint on available space, which is often the case for industrial sites and particularly so when the membrane must be retrofitted. It is also evident that the footprint incurred by the pumped sMBR configuration is significantly lower than all other configurations, a key motivating factor for its selection over the lower-energy immersed configurations if spatial constraints are particularly severe.

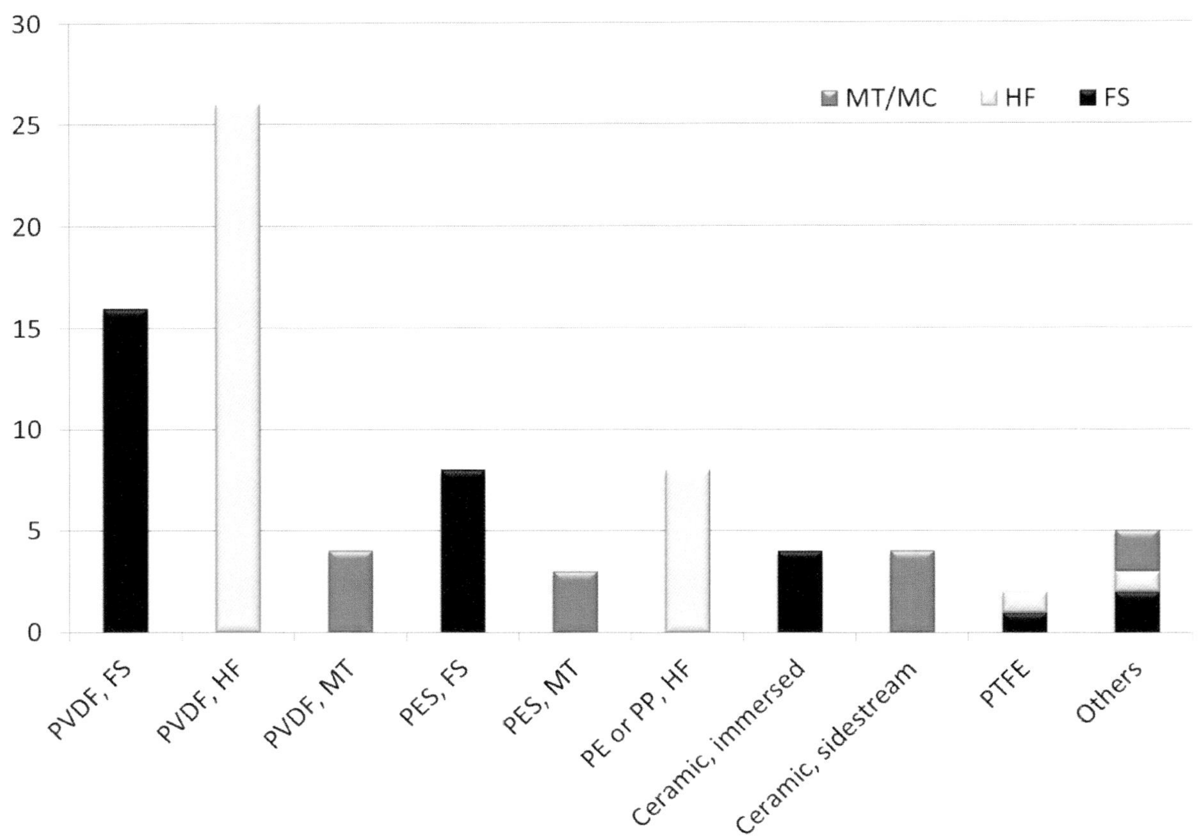

Figure 4-15 Total MBR membrane configurations

References

Marketsnresearch (2013). Global and China MBR Membrane Market 2013 Research Report, www.marketsnresearch.com/report/165260.php, accessed June 2014.

GWI (2012). Industrial Desalination & Water Reuse, Report, Global Water Intelligence.

Judd and Judd (2011), The MBR Book (Second Edition), Butterwoth-Heinemann, Lon/NY.

GWI (2014). Market profile: The low-pressure membrane market. Recovering from a low point, Global Water Intelligence, 45-47 (March, 2014).

5 Case studies

Three key parameters which characterise the performance of an MBR plant are the product water quality (and the related % target contaminant removal), the permeate flux and the energy demand. The flux then determines the fouling rate, or the rate of increase in transmembrane pressure (TMP) with time, which in turn determines how often the membrane is cleaned and the methods adopted for cleaning.

The primary product water quality determinant with respect to industrial effluents is COD (chemical oxygen demand). BOD (biochemical oxygen demand, which almost invariably refers to a test period of five days) and nutrient concentration – quantified variously as ammonia ($N\text{-}NH_4$), Total Kjeldahl nitrogen (TKN), total nitrogen (TN) and total phosphorus (TP) - may also be of importance. The permeate total suspended solids (TSS) concentration is normally negligible for an MBR, but colloidal solids manifested as the silt density index (SDI) may be important if the effluent is to be subsequently treated by reverse osmosis (RO) or nanofiltration (NF) for reuse. Depending on the reuse duty, further polishing using UV irradiation may be required.

The specific energy demand (*SED* or *E* in kWh per m^3 permeate) for membrane operation relates to the net permeate flux in $L/(m^2.h)$ (or LMH), the TMP (the ratio of the flux to TMP being the permeability in LMH/bar) and the energy consumed in applying shear, as well as the downtime for cleaning. Shear may be generated either through air scouring of immersed membranes, in which case the key parameter is the specific aeration demand with respect to membrane area (SAD_m in $Nm^3/(m^2.h)$), or through the crossflow velocity (CFV) for pumped sidestream membranes. Shear helps to sustain the flux whilst limiting the decrease in membrane permeability, as reflected in an increased TMP for constant flux operation. Permeability is also sustained by physical and chemical cleaning, the latter normally applied as a clean-in-place (CIP) usually based on sodium hypochlorite, which then incurs process downtime. The frequency of cleaning, and the protocols used, are therefore key operation and maintenance (O&M) parameters.

Technology performance tends to change with flow, and hence the difference between the average and peak daily flow (ADF and PDF respectively) is important and usually requires flow equalisation (EQ) using an appropriately sized buffer tank as pre-treatment. Other standard pre-treatment processes include coarse and fine screening (>5 mm and <3 mm respectively, with the fine screen rating being generally dependent on the MBR technology). DAF may be required if free (i.e. suspended) oil is present in the feed.

Biological treatment may be aerobic (Ae), anoxic (Ax) or anaerobic (An) depending on the prevailing conditions in the process tank. Biological nutrient removal (BNR) is normally achieved by a combination of these conditions (Section 2.3.3), recirculating the mixed liquor (or sludge) through the different tanks. The sludge must also be recirculated through the membrane separation tanks at a given rate, normally denoted as a ratio with reference to the feed flow Q. Biological conditions are otherwise controlled by the hydraulic and solids retention times (HRT and SRT respectively).

In the following sections, over 50 case studies are presented, based on information provided for a wide range of MBR membrane configurations (flat sheet, FS, hollow fibre, HF, and multi-tube, MT) and applications. Both immersed (iMBR) and sidestream (sMBR) technologies are included, with the latter encompassing examples of the air-lift (A-L sMBR) configuration. Contributors are specifically identified in the front section of this book (page iv). The sections are identified by application, as in Chapter 3, and the key data trends are summarised in Section 5.9.

5.1 Food and beverage

5.1.1 Canada Malting, Calgary, Canada

The Canada Malting Co. Ltd produces malt for the brewing, distilling and food markets nationally and internationally. It is Canada's largest malting company, producing approximately 450,000 tonnes of malt per year.

The effluent treatment process is required to reduce the load discharged to sewer; an MBR was selected for this duty primarily because of the smaller footprint incurred compared with alternative technologies. The plant receives an ADF of 4,240 m^3/d (6,060 m^3/d PDF) of malting process effluent. Effluent COD concentrations range from 1,260 to 4,460 mg/L and the temperature is generally between 18 and 22°C.

The water is equalised for 9.5 hours (under average daily flow conditions) and passed through a 1 mm rotary drum screen. Chemical dosing with urea is required for nutrient balancing. The water passes into the biological process tanks (1,000 m^3 anoxic, 7,710 m^3 aerobic) and recirculated through the 189 m^3 membrane tank at a recycle ratio of 4-5Q. The MLSS is held at 10,000 and 12,000 mg/L on average in the process and membrane tanks respectively. The HRT and SRT are 2.1 and 21 days respectively for the average flow, though in practice the SRT varies between 15 and 48 days according to the shift patterns.

The membrane tanks (Fig. 5-1) are fitted with three trains of three cassettes containing 48 x 34.4 m^2 *ZW500D™* modules (Section 4.2.4). The total membrane area of 14,849 m^2 equates to a net flux of 11.9 LMH on average, with the TMP ranging between 70 and 550 mbar. The membranes are air scoured at a rate of 0.095 $Nm^3/(m^2.h)$ and backflushed at 34 LMH for 30 s every 10 minutes. Maintenance cleaning with 300 mg/L NaOCl is applied twice weekly for 50 minutes, supplemented with a weekly citric CIP at 2,000 mg/L. Recovery cleaning by soaking for at least 6 hours in 1,000 mg/L NaOCl followed by 2,000 mg/L citric acid is applied twice yearly or as required.

The product water has COD, TKN and ammonia levels of 100, 4-8 and <0.5 mg/L respectively.

Figure 5-1 The three membrane tanks at Canada Maltings

5.1.2 Holland Malt, Eemshaven, the Netherlands

The plant at Holland Malt (1,400 m^3/d average flow, 65 m^3/h peak flow) in Eemshaven was the first turnkey MBR installation for maltings effluent treatment, the plant having come online in June 2005. The technology was selected for wastewater recovery and reuse, conserving 80% of the freshwater supplies. The treated water is reused for upstream processes following polishing with UV irradiation, the originally installed GAC having been found to be unnecessary.

Feedwater at 25-35°C passes from the 2 m^3 pump pit to a 1,500 m^3 buffer tank, providing 26 h of equalisation at the design flow rate. It then passes through a 1 mm screen before passing on to the biological tank, containing a 300 m^3 anoxic zone and a 1,730 m^3 aerobic zone (and hence 35 h HRT), where the MLSS is held at around 15,000 mg/L.

The sludge, at 15,000 mg/L solids concentration, is pumped through 4 lines of 6 x 27 m^2 Pentair *Compact 27* MT modules (Section 4.3.1.4, Fig. 5-2) at a feed pressure of 5 bar and a mean CFV of 3.5 m/s. The flux generated by the 648 m^2 of membrane area is generally between 110 and 125 LMH, substantially exceeding the design flux of 90 LMH and allowing one line to be on standby. Membrane chemical cleaning is through the application of 400 mg/L NaOCl and 1 wt% citric acid every 6 to 8 weeks.

The plant achieves >99.7% removal of BOD down to permeate levels below 5 mg/L, and >97% removal of the COD to below 40 mg/L. TKN and TP effluent levels are <5 and <1 mg/L respectively, the latter assisted by ferric dosing. It is estimated that the footprint of the plant at ~350 m^2 is just over one third that of a plant based on CAS and that sludge production and operational costs are respectively around 33% and 20% lower.

5.1.3 Obolon maltery, Ukraine

The installation of the plant at the Obolon maltery in Ukraine followed the success of the landmark plant at Eemshaven (Section 5.1.2), commissioned in 2005. The Obolon plant is based on the same Pentair pumped sidestream membrane technology (Section 4.3.1.4), and the treated water is reused for general purpose within the factory. The construction of the malt house was divided into two phases, the first phase, completed in May 2008, producing a wastewater stream of 1,500 m^3/d on average and the second, completed the following March, generating another 900 m^3/d, yielding 2,400 m^3/d in total. The effluent temperature is normally at 30-35°C and contains around 1,000 mg/L COD, 400 mg/L BOD and 50 mg/L TKN.

(a)

(b)

Figure 5-2 The Pentair sidestream MT modules at (a) Holland Malt, and (b) Obolon

The treatment process comprises sedimentation, equalisation and fine screening through two sets of screens prior to and following the buffer tank. The final screen is a rotating drum rated at

1 mm. The water then enters the biological treatment tanks, consisting of an anoxic and aerobic tank of 450 m^3 and 2,650 m^3 volume respectively (and thus 31 hours HRT at the average flow rate), where the MLSS concentration is held at 10,000-11,000 mg/L. Chemical dosing with ferric chloride is applied for phosphorus removal.

The sludge is recirculated through the membrane skid at 4 m/s CFV and applied pressures of up to 5 bar. The skid comprises 6 lines of 6 *Compact 27* Pentair modules (Fig. 5-2b), providing a total membrane area of 972 m^2 and thus a net flux of 103 LMH at the average flow rate. In practice fluxes of up to 140 LMH are attainable. The membranes are chemically cleaned monthly for two hours with 300 mg/L NaOCl and quarterly for four hours with 1 wt% citric acid by forward flushing with the reagent in both cases.

The process removes 94-97% and >99% of the COD and BOD respectively and leaves residual concentrations of 30-50, 1-3, <5 and ~2 mg/L of COD, BOD, TKN and P respectively. The calculated OPEX was between 0.3 and 0.4 EUR/m^3 when reviewed in 2009.

5.1.4 Polar brewery, Caracas, Venezuela

Polar beer is the largest and best known brand of beer in Venezuela, the company having been established in 1941. The plant at Caracas generates 2,000 m^3/day of effluent, between 30 and 35°C, of which up to 70 m^3/h (~1,700 m^3/d) is reused following treatment by MBR and RO. The Pentair-based A-L sMBR technology (Section 4.3.1.4, Fig. 5-3a) is similar to that originally applied at Ootmarsum for municipal wastewater treatment (Fig. 5-3b), reconfigured to reduce the footprint. The plant has been in operation since Dec 2008, and on-site reuse duties for the recovered water include CIP water preparation, cooling and steam-raising.

(a) (b)

Figure 5-3 The Pentair *Airlift™* technology: (a) Polar, Dec 2008, (b) Ootmarsum, Oct 2007

The MBR feedwater, at around 300 mg/L COD, derives from the outlet of a Paques anaerobic digester, and undergoes equalisation in an existing tank prior to screening using a 2 mm Amiad basket strainer. It then enters a 2,000 m^3 aerobic tank held at 10,000–11,000 mg/L MLSS. The sludge is recirculated through a sidestream Pentair *Airlift™* membrane system comprising two skids of 30 *Compact 27* Pentair modules (Fig. 5-3a), giving a total membrane surface area of 1,980 m^2 and so a net flux of 42 LMH at the average feed flow rate. The sludge is pumped at a crossflow of 0.5 m/s assisted by air-lift at around 0.3 $Nm^3/(m^2.h)$. The membranes are maintained by monthly cleaning for 2 hours with 300 mg/L NaOCl supplemented with quarterly cleans for 4 hours with 1 wt% citric acid.

The process removes >93% and >83% of the COD and TKN respectively and leaves residual concentrations of <20 and <10 mg/L of each. The calculated OPEX for the plant was between 0.3 and 0.4 EUR/m^3 when determined in 2009.

5.1.5 Shepherd Neame Brewery, Faversham, UK

Shepherd Neame, based in the market town of Faversham in Kent, is the UK's oldest brewery. The key challenge posed by the effluent treatment project was the small available footprint for the plant, which precluded the use of both conventional technologies and immersed MBR systems operating at lower MLSS concentrations. The new treatment plant, designed by Aquabio and commissioned in October 2013, employs RO and UV post-treatment to recover 40% of the effluent for reuse in CIP operations and boiler feedwater.

The plant is designed for 700 m^3/d average daily flow of wastewater at a temperature of 25-40°C (30°C on average). The design includes as pre-treatment a DAF, a 0.25 mm rotary drum fine screen, and flow balancing providing ~8h of EQ. Nutrient balancing is employed, as well as occasional antifoaming agent dosing. There is also screening of the RAS using a simple 1 mm basket strainer.

The core biological treatment is based on the Aquabio low-energy *AMBR LE*™ concept (Section 4.3.3.1). The 1,480 m^3 aerobic tanks provide a ~2-day HRT, and the mean MLSS concentration of 12,000 mg/L is sustained by a ~27-day SRT. The membrane skid comprises 2 loops of 6 modules, with 2 modules per line, yielding 640 m^2 total area and a net design flux of 46 LMH which is regularly exceeded in practice. The membranes are backflushed for 30 s every 30 minutes, supplemented with a 3-hour CIP with NaOCl every other month to maintain membrane permeability. Membrane permeation incurs an energy demand of 0.45-0.6 kWh/m^3, the demand decreasing at lower fluxes where the use of VSD (variable speed drive) pumps permits reduced pumping.

There have been some issues with RO fouling at this site, resolved by appropriate chemical conditioning and CIP protocols. Both the MBR UF membrane flux and the final water quality achieved are above the design expectations, BOD and ammonia being below 20 and 5 mg/L respectively.

5.1.6 Arla Foods, Vimmerby, Sweden

The Arla Foods plant at Vimmerby in Sweden is a whole dried milk production facility, which began operations in August 2003. Since June 2012 there has been a Grundfos *BioBooster* MBR plant (Section 4.3.2.1) operating at the site, fed with a mean effluent daily flow of 400 m^3/d (29 m^3/h peak hourly flow), which was installed to meet the strict consent for discharged P.

The effluent temperature ranges between 20 and 35°C (27°C on average) and has COD and TN concentrations of 3,000-4,700 and 200-350 mg/L respectively (3,500 and 280 mg/L on average). Effluent is screened using a 0.6 mm rotary drum screen before passing on to the biological process tanks. The biological treatment is configured for BNR and consists of anaerobic, anoxic and aerobic tanks of 180, 280, 260 m^3 volume respectively (and hence providing a mean overall HRT of 43 hours), with the sidestream membrane separation skid adding a further 8 m^3. Recirculation through the membrane modules is at 3-4Q, maintaining the membrane module MLSS concentration at 15,000-25,000 mg/L for a process biotank concentration of 8,000-14,000 mg/L at the SRT of 20-30 d.

Membrane filtration is based on a bank of 16 tubular modules each offering a membrane area of 34 m^2 from 180 rotating ceramic discs. The net flux is therefore 32 LMH at the mean daily flow rate, sustained by an applied pressure of 500 mbar. The discs are backflushed every 10 minutes for 15 seconds and maintenance cleaned for 36 minutes weekly with 2 wt% hypochlorite followed by 1 wt% combined acetic and nitric acid.

The plant achieves a consistently high effluent quality of <30 mg/L COD, <4 mg/L TN and <0.3 mg/L TP, the latter attained without necessitating supplementary chemical dosing.

5.1.7 Basic American Foods, Blackfoot, US

The Basic American Foods plant at Blackfoot, Idaho treats effluent from a potato processing plant, the site of possibly the first potato dehydration facility in the US. An MBR was selected to provide a treated water quality high enough to be used directly for crop irrigation, demanding total nitrogen removal, and also within the factory itself for fluming raw material. The plant (Fig. 5-4a) treats a maximum daily flow of 6,400 m^3/d and was installed in 2002 when the primary contractor was Zenon Environmental Corp (now GE Water and Process Technologies) and the primary consultant was Donohue and Associates.

The wastewater temperature varies between 20 and 40°C and has mean COD and TKN concentrations of 240 and 94 mg/L TKN respectively. The effluent undergoes screening via two pairs of shaker tables, one pair rated at 3.4 mm and the other at 1.7 mm. The effluent is pre-treated in a cylindrical primary clarifier (666 m^3, 2.5 h HRT at the maximum daily flow) before passing on to the MBR consisting of the anoxic, aerobic and membrane tanks, respectively having volumes of 4,550 m^3, 9,100 m^3 and 227 m^3 yielding a total HRT of around 52 hours at peak flow. The MLSS concentration in the aerobic and membrane tanks is 8,000-12,000 and 10,000-15,000 mg/L respectively, the recycle ratio being 5Q.

Membrane separation comprises 3 trains of 10 cassettes each fitted with 22 *ZW500C™* modules (Fig. 5-4b) having a membrane area of 23.2 m^2, giving a total membrane surface area of 15,312 m^2. The 12-minute filtration cycle includes 35 s of backflushing and 20 s of relaxation to sustain a design flux of 17.5 LMH. Membrane air scouring was originally applied using standard intermittent 10:10 aeration (10 s on, 10 s off, Section 2.3.5.2), but after monitoring membrane performance it was possible to reduce membrane air scouring to 10 s on, 20 s off – equating to a SAD_m of 0.22 $Nm^3/(m^2.h)$ on average. Maintenance cleaning is carried out weekly by applying 100 mg/L of sodium hypochlorite for 45 minutes, sustaining a TMP of 70-550 mbar. During the 12 years of operation no more than five recovery cleans have been required, these comprising soaking in 2,000 mg/L citric acid for 2 hours. All but three of the cassettes were replaced after 10 years: the mean membrane life was thus double the warrantied life of 5 years.

(a)

(b)

Figure 5-4 (a) The membrane building and (b) a membrane cassette at the Basic American Foods WwTP

Membrane and biological aeration incur energy demands of 0.39 and 0.92 kWh/m^3 respectively, with sludge transfer and permeation demanding a further 1.71 and 0.27 kWh/m^3. The total energy demand is therefore around 3.3 kWh/m^3. The plant sustains a consistently high effluent quality of ~13 mg/L COD and ~4 mg/L TN.

5.1.8 KMC, Brande, Denmark

KMC, founded in 1933, is a leading producer of starch and other potato based products. The company has three factories located close to the potato farmers in Denmark that produce native potato starch, two being modification plants and the third for potato granules and flakes production. The 1,000 m^3/d capacity MBR plant installed at its Brande site was the result of operational problems with the existing CAS plant in the form of a spirochete bloom which caused sludge bulking. Membrane separation was fitted in Nov 2007 following successful pilot testing to compare FS and sidestream technologies. The contractor for the installation was KD Maskinfabrik A/S.

The effluent passes through 1 mm screening before nutrient adjustment with urea and phosphoric acid. It then passes on to the 6,900 m^3 biotank (providing 6.9 d HRT at the design flow) where the MLSS is held at 6,000 mg/L, compared with 8,000 mg/L in the 60 m^3-capacity membrane tank through which the sludge is recirculated at a ratio of 4Q. The SRT is around 25 days. Membrane separation comprises 4 trains of 4 x 308 m^2 double-deck Alfa Laval units (*MFM200*, Section 4.1.1), offering a total membrane area of 4,928 m^2. At the design flow rate of 1,000 m^3/d the corresponding net flux is 8.5 LMH. The membranes are air-scoured at a rate of 0.3 Nm^3/(m^2.h) and relaxed for 2 minutes every 10 minutes. A two-hour monthly maintenance clean is conducted with 1,000 mg/L hypochlorite followed by HCl at pH 1.8. Under these conditions the maximum TMP recorded is around 80 mbar, with permeabilities generally between 400 and 600 LMH/bar.

The membranes, warrantied for 3 years, were changed after 6 years of operation (complete module replacement) following a deterioration in effluent quality resulting from membrane mechanical integrity issues. This was attributed to the aggressive cleaning conditions employed (monthly soaking in HCl at pH 1-1.5), which have subsequently been adjusted to pH 1.8. The cleaning is demanded by the unusually high calcium carbonate concentrations in the effluent, which also accounts for the low net flux: only three of the four membrane tanks are online for much of the time. The COD is reduced from an average 8,000 mg/L (maximum of 36,000 mg/L) down to 50-200 mg/L, with TKN and ammonia reduced to <5 and <1 mg/L respectively.

(a)

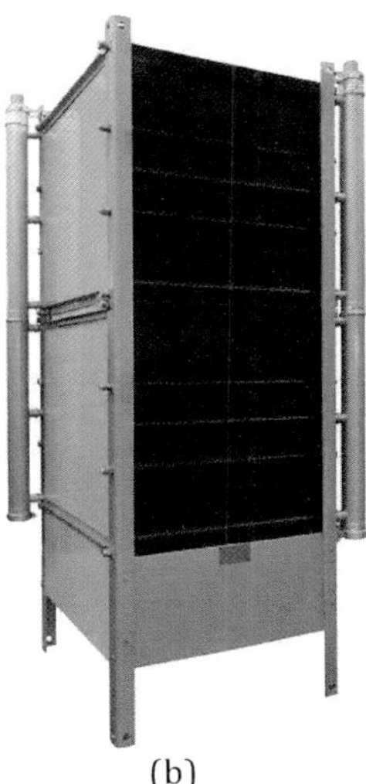

(b)

Figure 5-5 (a) The membrane tanks, and (b) the Alfa Laval *MFM200* unit at the KMC site

5.1.9 Novidon, Belgium

Novidon, part of the agro-industrial group Royal Cosun (annual turnover ~2b EUR in 2013), creates and manufactures potato starch-based products for applications across a variety of industrial sectors which include paper, textile and oil and gas. The effluent treatment plant receives an average flow of 240 m^3/d ranging in temperature between 6 and 36°C (18°C on average).

Primary clarification at a mean surface overflow rate of ~20 m/d is used to reduce levels of suspended starch using a polymeric flocculant, with EQ for 15 and 7.5 hours in two separate tanks. The effluent then enters a single 1,480 m^3 aerobic tank (6.2 d HRT), where nutrient balancing is applied. The MLSS is maintained at 5,300 mg/L by an SRT of 33 d and transferred to the membrane separation process.

The membrane technology used is the Berghof *BioairDS* (Section 4.3.3.2) providing 320 m^2 of membrane area as vertical sidestream MT modules of 53.4 m^2 each. These are air-scoured at a net rate of 24 Nm^3/h in total (and hence SAD_m = 0.075 $Nm^3/(m^2.h)$) supplementing the liquid crossflow of 0.5 m/s. The membranes are backflushed at a rate of 150-200 LMH every 30-45 minutes for 30-45 s. These conditions sustain a net flux of 31 LMH and a TMP below 300 mbar. The membrane is cleaned for ~2 hours monthly with acid and twice monthly with NaOCl, both at 100 mg/L.

The product water has COD, TN, TKN and TP levels of 20, 3, 2 and 0.5 mg/L respectively, P being removed primarily by assimilation and only occasionally requiring coagulant dosing. The treated water is reused for non-food applications within the industrial process, the surplus being discharged to a surface water.

5.1.10 Brodheadsville, PA, US

An MBR was selected for the beverage production plant Brodheadsville, PA, primarily on the basis of a lower investment cost compared with that of the alternative anaerobic treatment technology also considered. A consistently high treated water quality is required for discharge to the receiving water body. The MBR technology was retrofitted to the existing assets, which featured a large aeration tank as well as EQ and sludge tanks, and has been in operation since April 2013.

The effluent – containing 250 mg/L TSS, 7,000-15,500 mg/L BOD and 9,000-20,000 mg/L COD – varies between 10°C and 32°C in temperature and is mildly acidic (pH ~ 5.5). The plant has a design capacity of 265 m^3/d (14.4 m^3/h peak hourly flow). The wastewater is passed through a 6 mm screen – the main coarse suspended solids present being stray bottle caps, and pH-adjusted before flow balancing for 32 hours prior to entering a DAF. It is then nutrient-adjusted using ammonium sulphate and potassium phosphate and passed on to the aerobic tank (970 m^3), from which the sludge is recirculated through the 57 m^3 membrane tank at a recycle ratio of 6Q. The HRT associated with the tankage is thus in the region of 4 days, providing ample EQ while also minimising sludge production. The SRT of 23 d sustains an MLSS of 12,000-14,000 and 14,000-16,000 mg/L in the process and membrane tanks respectively, the *F:M* ratio being 0.17 d^{-1} with respect to COD.

The two membrane tanks are each fitted with 6 double-deck MaxFlow *U70-003* membrane modules (Section 4.1.13) of 70 m^2 per module, providing 840 m^2 of total membrane area and a corresponding net flux of 13.1 LMH at the 265 m^3/d design flow. The membranes are air-scoured at a rate of 0.35 $Nm^3/(m^2.h)$ to maintain the TMP between 7 and 240 mbar. Chemical cleaning employs 0.5 wt% NaOCl and is applied as required, the first one after 8 months of operation. The plant produces effluent of below 40 mg/L COD.

The total project budget was a little over \$1.4m, including all equipment, tank and building refurbishment, wiring, engineering and construction. The total power available at the site is limited to 400 A at 480 V (3-phase, 60 Hz). Whilst a larger blower was required for the EQ, sludge and aerobic tanks, it is operated via a VSD along with modulating valves and air flow meters. This has resulted in a lower than anticipated *SED*, since the aeration is optimised according to demand: the current usage is often less than half of the total available capacity.

(a) (b)

Figure 5-6 (a) Aerobic tank and (b) MaxFlow membrane module at Brodheadsville

5.1.11 Food factory, Machida, Japan

The 900 m^3/d (62 m^3/h peak flow) iMBR at a food preparation site in Machida, Tokyo, was commissioned in July 2004. Wastewater treatment at this site was originally based on SBR technology, but an upgrade was required due to increased effluent flow from the factory. Spatial restrictions at the site meant that MBR technology was considered the most viable option.

The wastewater passes through a 2 mm vibrating screen to a 950 m^3 aerobic/membrane tank (i.e. ~25 h HRT): there is apparently no denitrification requirement at this site. The mean MLSS in the process tank is 10,000 mg/L, maintained by an SRT of 32 d. Membrane separation is provided by two trains of 12 skids, each containing 4 Asahi Kasei *MUNC-620A* MF 25 m^2 membrane modules (Section 4.2.1), yielding 2,400 m^2 of membrane area in total. The membranes are operated on a cycle 9 minutes filtration and 1 minute of backflushing (at 16 LMH), sustaining a net flux of 16 LMH with a mean TMP of 0.15 bar. The membranes are scoured at a rate of 0.20 Nm^3/h air per m^2 membrane. Maintenance cleaning with 1,000 mg/L NaOCl is conducted twice monthly for 1.5 hours, and this is supplemented with recovery cleans using 3 g/L NaOCl or 1 wt% oxalic acid every two years or as required.

The treated effluent contains <5 mg/L BOD_5, <30 mg/L COD and has non-detectable ammonia. The membranes were replaced in November 2013, exceeding their 5-year warrantied life by more than 4 years.

5.1.12 Food processing plant, Taipei, Taiwan

The MBR plant for treating food processing effluent was installed in May 2013 by Ecologix to upgrade the existing plant, and is based on an *EcoMem* compact MBR.

The existing process had pre-treatment based on a 1 mm rotary drum screen followed by flocculation by polymer dosing and DAF clarification. This was followed by a CAS process designed to treat 400 m^3/d. An upgrade was required to meet an increased wastewater flow of 600 m^3/d, which the client wished to achieve without installing further concrete tanks.

The package plant provided for the upgrade was the Ecologix *EcoMem-600PKG* (Fig. 5-7a) containing a 12,000 mm long x 2,000 mm wide x 2,400 mm high membrane tank constructed of SS 304. The tank is fitted with 12 *EcoPlate™-120S* cassettes of 103.2 m^2 each (yielding 1,238 m^2 total), each cassette housing 120 *EK-08* panels (Section 4.1.5). Thus, at the design flow of 25 m^3/h, the flux is around 20 LMH. The membranes are scoured at a rate of 0.52 $Nm^3/(m^2.h)$ (rated at 3 m head). The MBR permeate is directed to an RO stage which provides water for the fluming of vegetables.

The wastewater is delivered to the existing aeration tank and continuously recirculated at a ratio of 2Q through the membrane tank via duty/standby sludge return pumps of 50 m^3/h capacity each. Two level switches in the membrane tank control the supply and MBR permeate pumps, the latter comprising two self-priming pumps each of 25 m^3/h capacity (duty/standby). The MLSS is kept within the range between 8,000 and 12,000 mg/L.

The MBR permeate produced has COD and BOD levels below 50 and 5 mg/L respectively, from an influent water containing 1,200 and 800 mg/L mean concentrations of COD and BOD. The package plant has successfully provided a plug-and-play solution for capacity expansion without requiring costly construction.

(a) (b)

Figure 5-7 (a) The *EcoMem* compact MBR installed at the site in Taipei, and (b) the *EcoMem*. Inset: permeate and sludge samples from the site

5.1.13 Kanes Foods, Evesham, UK

Two separate plants have been installed at Kanes Foods in Worcestershire, UK, for food effluent recycling using the pumped UF MT sMBR technology of Aquabio, a Freudenberg Filtration Technologies company, with a two-stage RO and UV fitted downstream. The original 815 m^3/d plant, operational since 2001, is based on the company's *AMBR*™ system (Section 4.3.3.1). The newer plant, commissioned in February 2010, has a capacity of 1,435 m^3/d. The proportion of the MBR feed recovered is ~75% and ~42% for the original and newer plants respectively, the RO permeate being blended with mains water for reuse within the factory. Both plants employ DAF treatment for fine vegetable solids removal followed by fine screening (0.5 or 0.25 mm for the original and newer plant respectively) with ~27 h EQ and a biotreatment HRT of 13-15 h. Both operate at 10,000-12,000 mg/L MLSS for most of the time, equating to an *F:M* ratio of around 0.13 kgCOD/kgMLSS day at the normal feed COD concentrations of <1,000 mg/L.

The original *AMBR*™ classical pumped sMBR plant is fitted with 4 loops of modules providing 324 m^2 total membrane area (Fig. 5-8a), and thus a peak flux of 105 LMH if all loops are operational. In practice an average flux of 153 LMH (normalised to 25°C) is attained, sustained by the CFV of 3-4 m/s and elevated feed pressure, permitting duty/standby operation. The *AMBR LE*™ membrane skid comprises 4 loops of 6 modules (Fig. 5-8b), with 2 modules per line, yielding 1,282 m^2 total area and a net flux of 47 LMH. The skid is operated at a reduced pressure and CFV compared with the *AMBR*™ plant resulting in an *SED* of ~0.5 compared with ~1.5 kWh/m^3. Significant process flexibility is offered in the *AMBR LE*™ system by the use of VSD recirculation pumps and optional permeate pumping to respectively control the TMP and flux. The membranes are also backflushed for a total of 8 minutes in each hour, maintaining the membrane permeability.

(a) (b)

Figure 5-8 The membrane banks at Kanes Foods (a) the original four banks from 2001 and (b) the new *AMBR LE*™ bank.

The plants have performed consistently well in terms of biological treatment, membrane performance and final reuse water quality (40–100 µS/cm). The MBR permeate generally contains mean TSS, BOD and COD values of 4, 7 and 17 mg/L respectively from a feed COD concentration of up to 1,000 mg/L, and the RO achieves an overall recovery of 75–80%. Occasional reductions in MBR membrane flux have been linked to poor biomass health which has been rectified by closer management of the process.

5.1.14 Esmeralda, meat processing, Lima, Peru

Esmeralda Corp supplies agro-processing, meat and seafood products for the local and international market. The effluent generated contains COD (550-5,800 mg/L), BOD (300-3,600 mg/L) and FOG (40-750 mg/L).

The MBR plant treats combined food processing and sanitary wastewater flowing at 800 m^3/d, and has been operational since 2011. Wastewater between 15°C and 25°C is pre-treated using a 1 mm rotary drum screen before flow balancing in a 400 m^3 tank (providing ~12 hours EQ) and passing on to a DAF. The clarified DAF effluent then enters the 400 m^3 aerobic tank in which the sludge is maintained at 4 g/L MLSS. This sludge is recirculated through four membrane tanks, two of 50 m^3 capacity and two of 60 m^3, where the sludge is concentrated to around 10 g/L. The total HRT at the design flow is 19 h.

Membrane separation is conducted using 18 Toray *MEMBRAY® TMR140-100S* single-deck modules (Section 4.1.22), each fitted with 100 x 1.4 m^2 panels, providing a total membrane surface area of 2,520 m^2. 4, 6 and 8 modules are respectively placed in the first, second and third tanks, with a further 8 modules in the fourth tank to provide additional capacity. The membrane operating cycle comprises 9 mins filtration and 1 min relaxation, the net flux being 13.2 LMH. Air scour is applied at an average rate of 0.43 $Nm^3/(m^2.h)$, and the TMP then maintained at 50-200 mbar. Chemical cleaning is conducted every 4 months by applying a 6 h soak in 2,000 mg/L NaOCl, supplemented with a quarterly 8-12 h soak in 4,500 mg/L citric acid.

The site includes a UASB reactor for pre-treatment of concentrated wastewater from fish meat processing (24,000 mg/L COD, 13,000 mg/L BOD) and an RO array for further purification of the treated water (100 m^3/d, averaging <30 mg/L COD in the MBR permeate) to allow its reuse for boilers and condensers. The MBR permeate is otherwise used for turf grass irrigation and cleaning the WwTP. 100% reuse of the wastewater is accomplished at this site.

5.1.15 Coosur, Jaén, Spain

The Coosur plant at Jaén in Spain generates an average daily effluent flow of 100 m^3/day (120 m^3/day peak) from olive oil production. A DAF process, pH-adjusted from 3.5 to 6.5 and dosed with coagulant and flocculant, is installed upstream of the MBR. The COD and TSS of the DAF influent is 4,500 mg/L and 700 mg/L on average respectively, with a conductivity of more than 4,000 μS/cm. The contractor and operator of the plant is Aqualia.

The project entailed the retrofit and upgrading of an existing SBR process plant to allow an increased treatment capacity and the opportunity to reuse the treated water. The *Likuid-CBR®* filtration skid was thus coupled to the existing biological reactor to provide an effluent free from suspended solids and bacteria and thus able to be post-treated by RO to generate a product water for reuse in industrial cleaning operations. The surplus effluent is discharged to a nearby surface water. A decanting centrifuge is used for sludge dewatering, prior to it being tankered off site.

The plant operates at an MLSS concentration of 8,000 mg/L, a jet aerator being used to maintain the DO levels in the sludge at >2 mg/L. The sludge is pumped at a feed pressure of 4.5 bar, and the system then operates with a nominal and maximum TMP of 1.8 and 2.5 bar respectively. The activated sludge is recirculated with a CFV of 3 m/s through a skid (Mode no. *CBR L3706*) fitted with 6 vertically-aligned Likuid Nanotek modules (Section 4.3.2). The skid (Fig. 5-9) provides 54.6 m^2 membrane area and incurs a footprint of 3,900 x 2,300 mm (9 m^2). The net flux is thus 83 LMH, sustained by chemical cleaning 3-5 times/year with NaOH and HNO_3 for alkaline and acid cleaning, respectively. The MBR permeate COD is below 50 mg/L, representing a removal of more than 96%, with a low SDI.

5.1.16 Chicken slaughterhouse, Bona Avis, Ianca, Braila district, Romania

The 730 m^3/d capacity effluent treatment plant at the Bona Avis chicken slaughterhouse was installed in 2009 to meet discharge regulations. The wastewater has a feed COD, BOD, TN and TP levels of 4,800, 2,400, 290 and 80 mg/L, along with suspended solids levels of 1,720 mg/L.

The initial stage of the treatment process is the mechanical separation of extraneous solids materials using a bank of three 0.5 mm rotary fine screens. Solids retained are discharged to a sludge tank and conditioned with lime before being pneumatically pumped to a filter press for dewatering, in preparation for incineration. The screened water then passes on to an aerated 543 m^3 buffer tank to provide 17 hours of flow and load EQ.

Figure 5-9 The membrane skid at Coosur

There then follows a physicochemical treatment stage comprising a DAF unit (pre-dosed polymeric flocculant) operating at a loading rate of 40 m/h. This retains the bulk of the suspended solids and fats, the sludge waste being fed to a heated filter press. The clarified effluent then passes on to the biological treatment stage, comprising a 434 m^3 anoxic tank (equipped with mechanical mixers) and an 1,842 m^3 aerobic tank, the total HRT thus being in the region of 3.1 days at the 730 m^3/d flow capacity.

The biotank MLSS is recirculated through the membrane tank at a recirculation ratio of 6Q. The membrane tank is fitted with 4 MICRODYN-NADIR *BIO-CEL® BC400* membrane modules (Section 4.1.16), distributed between two cylindrical PP tanks. The 4 modules provide a total membrane area of 1,600 m^2 which, at the stated flow capacity, yields a net flux of 19 LMH. The flux is maintained by air scouring at a rate of 0.2 $Nm^3/(m^2.h)$ combined with backflushing every 9.5 minutes for 30 seconds and a monthly CIP. The CIP comprises a 2h soak in 250 mg/L hypochlorite every 28 days and intensive yearly cleaning with 5,000 mg/L citric acid. The membrane tank MLSS concentration is 10-12 g/L (according to supplier's specifications), controlled by a TSS sensor. When this concentration is reached, one TSS sensor signals the pumps to discharge WAS to the storage tank.

The treatment scheme is designed to reduce the pollutant load to permitted values for discharge, as regulated by NTPA 001/2005. The plant has consistently achieved a product water quality of 125, 25, 15 and 2 mg/L of COD, BOD, TN and TP, equating to >97% COD removal.

5.1.17 Arla Dairy, Aylesbury, UK (anaerobic MBR)

The 400-700 m^3/d milk processing effluent treatment plant at the Arla Dairy in Aylesbury in the UK is based on the Veolia Water Systems' *Memthane* anaerobic sMBR process (Section 4.3.4). In this case anaerobic treatment was pre-selected by the client, whose stated ambition for the new dairy was for the installation to showcase sustainable development, applying advanced effluent treatment process technologies incorporating renewable energy.

The wastewater, at 14-27°C (22°C average) and having a mean COD of 11,700 mg/L, is coarse-screened, chemically-conditioned, and equalised for 16 h (at the maximum flow) before passing through a heat exchanger and on to the biotank. The MLSS is held at a concentration of 15,000 to 30,000 and recirculated through a bank of 6 loops of 7 modules at a CFV of 1.5-3.5 m/s. At a

TMP of 0.1-0.5 bar the flux generated is in the region of 12-20 LMH. The membranes are cleaned with monthly citric acid and hypochlorite CIPs.

The effluent is post-treated with flash aeration to remove residual COD, achieving more than 99.5% removal of the feed COD overall.

5.1.18 Potato processing wastewater and tequila stillage pilot trials, anaerobic MBR

A 4-month pilot-scale study of the treatment of effluent from a Canadian french-fry production plant, was conducted by the University of New Brunswick in collaboration with the company ADI Systems Inc. (Section 4.1.24) in Canada, who designed the plant (Singh et al., 2010) and have installed a number of full-scale AnMBR plants. The plant comprised a 2.0 m^3 completely mixed anaerobic tank in conjunction with a 1.3 m^3 membrane tank.

Wastewater at ~38,000, ~12,300 and 17,400 mg/L COD, BOD and TSS concentration was fed into the 35°C reactor at a rate of 0.3-2.2 m^3/d and an HRT generally between 3.5 and 14 days. The OLR was steadily increased over the course of the first 10 weeks of the trial and the plant then operated at an *F:M* ratio of 0.09-0.18 gCOD/(gMLVSS.d) and an MLSS concentration of 41,000 mg/L MLSS for the remainder of the study. Under these conditions the SRT was 80 d and the sludge yield 0.083 gTSS/gCOD, demonstrating significant digestion of the influent potato suspended solids.

The membrane tank was fitted with 25 x 0.8 m^2 *510 ES* Kubota FS membrane panels (and thus a total membrane area of 20 m^2 providing a net flux of 0.5-2 LMH) and the sludge recirculated through it at a rate of 5Q. Membranes were continuously scoured with the generated biogas throughout. The TMP remained below 0.04 bar at all times other than when the flux was raised to >4.2 LMH, at which point the TMP increased rapidly to 0.1 bar and a 10% citric acid clean was employed to recover the permeability. During the stabilised period a flux of 3.1 LMH was sustained without fouling.

The plant achieved >99% COD removal (360 mg/L permeate concentration) and >99.5% throughout the stabilised period of operation.

The outcomes of a further trial on tequila stillage effluent, also conducted by ADI Systems Inc., was reported at the same time as the above trial (Grant et al., 2010). In this case the pilot plant was operated for 105 days between Nov 2008 and May 2009 and challenged with a feed containing 35,000-70,000 mg/L COD (56,900 mg/L on average) and 14,750 mg/L TSS. The biological operating conditions comprised 12.4 d HRT, 70 d SRT and 37°C. These conditions provided influent SS degradation of ~70%, based on a sludge yield of 5% of the COD removed (i.e. 0.05 kgTSS/kgCOD removed).

The same FS membrane cassette of 25 x 0.8 m^2 Kubota *510 ES* FS membranes was used as for the potato processing effluent. For the tequila stillage effluent trials the permeate was recirculated at 2 m^3/d through the membranes to extend the HRT without reducing the flux. Under such conditions a flux of 4.2 LMH was sustained for the entire 105-day period without a CIP following an initial clean with 10 wt% citric acid.

5.2 *Petroleum and heavy industries*

5.2.1 Sinopec Guangzhou, China

Sinopec Guangzhou Company is an oil refining and ethylene production company with annual capacities for crude oil processing and ethylene production of 13.2 m and 0.22 m tons respectively. The company, based in Southern China, has invested 1,600 m RMB (~250 m USD) in total for constructing environmental protection facilities, which include its MBR for treating the refinery effluent.

The MBR, designed and installed by NOVO Envirotech (Guangzhou) Ltd, treats 4,800 m^3/d of effluent whose temperature varies between 20 and 40°C. The plant has been operating since March 2008. The plant is designed to remove at least 80% of the COD, 90% of the BOD and 95% of the ammonia to leave residual concentrations of <45 mg/L, <2 mg/L and <1 mg/L respectively. Mineral oil is removed to below 1 mg/L and phenol and sulphide to below the limit of detection.

Pre-treatment comprises equalisation combined with two stages of flotation and sedimentation and an oxidation ditch to reduce the load onto the downstream MBR (Fig. 5-10). The plant is based on anoxic, aerobic and membrane tanks which are respectively 900 m^3, 1,600 m^3 and 500 m^3 in volume, with transfer between the membrane and aerobic tank at a rate of 2-3Q. Operation is at 15 HRT and 30-90 d SRT respectively, providing an MLSS concentration of 2-4 g/L in the process tank and 3-8 g/L in the membrane tank, reflecting the comparatively low recirculation ratio. Membrane separation is via four trains (Fig. 5-11) each fitted with five cassettes of 56 20 m^2 *SMM-1520* Memstar modules (Section 4.2.11), providing a total membrane area of 22,400 m^2 (and hence a net flux of 8.9 LMH). These are aerated at a SAD_m of 0.07 $m^3/(m^2.h)$ and operate with 9 minutes filtration and 1 minute relaxation. Weekly maintenance cleaning with 500 mg/L NaOCl is supplemented with recovery cleans using 1,000-2,000 mg/L NaOCl with 0.1 wt% NaOH and surfactant once or twice yearly.

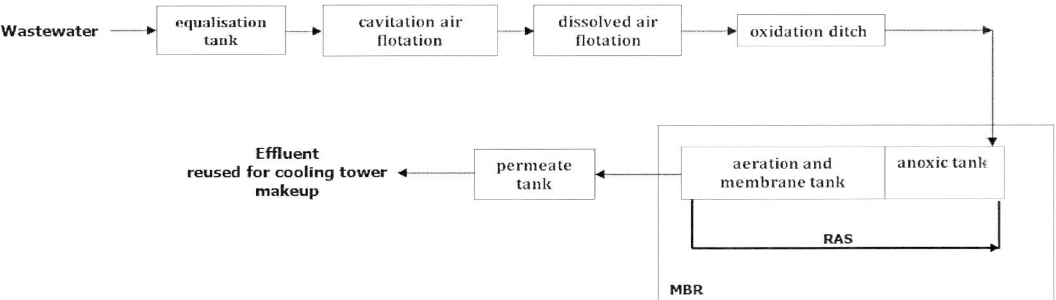

Figure 5-10 The effluent treatment scheme at Sinopec Guangzhou

Figure 5-11 The membrane tanks at Sinopec Guangzhou

The plant has generally performed well. Although mineral oil (>10 mg/L) in the influent stream had caused severe fouling on occasions, this was addressed by the development of a bespoke cleaning reagent by the technology supplier Memstar, which has satisfactorily recovered permeability. The *SED* is comparatively low at 0.6 kWh/m^3, with 73% accounted for by aeration and 0.06 and 0.1 kWh/m³ respectively consumed by sludge pumping/dewatering and permeate pumping operations.

5.2.2 Formosa, Yunlin, Taiwan

Formosa Petrochemical Corporation is the second largest company for oil refining in Taiwan, producing fuel products and petrochemicals. The company owns gas stations, through subsidiary Formosa Oil, and sells electricity and steam from its co-generation plants. The company's refinery and petrochemical plant at Yunlin County discharges an effluent mean flow of 20,000 m^3/d to an MBR plant (Fig. 5-12), installed in 2002. The permeate (~50 mg/L COD) is used for utilities (cooling) and general purpose (cleaning).

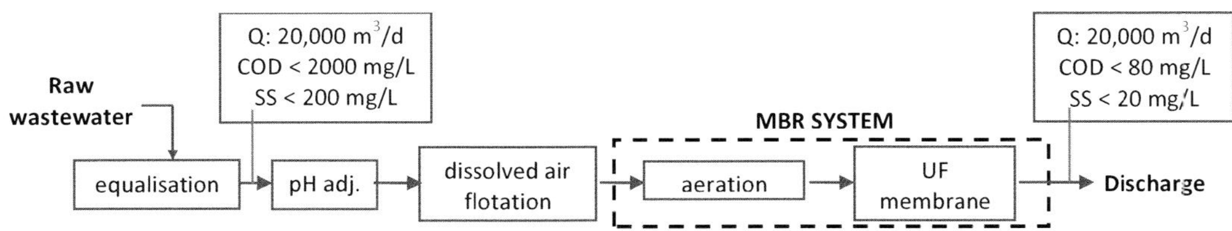

Figure 5-12 The treatment scheme at Yunlin

Wastewater at 20-30°C and containing around 1,000 mg/L COD is equalised, pH-adjusted and then DAF treated before being transferred to the biotank (Fig. 5.13a), where the MLSS is held at 3,500 mg/L. The sludge is recirculated, at 4Q, to the membrane separation stage (Fig. 5.13b) comprising four trains of eight cassettes fitted with 48 GE *ZW500D™* modules (Section 4.2.4) making 1,536 modules in total. The total membrane area of 52,531 m² provides a net flux of 16 LMH on average, with instantaneous fluxes between 16.5 and 18 LMH and the TMP ranging between 70 and 500 mbar. Maintenance cleaning with 500 mg/L NaOCl is applied once or twice weekly for 40 minutes, and a supplementary 6 h recovery clean with 1,000 mg/L NaOCl followed by 2,000 mg/L citric acid applied 2-4 times a year.

The plant has run without major upsets with recovery cleans applied when demanded by the permeability decline, and the membrane life has exceeded six years as of June 2014.

(a) (b)

Figure 5-13 (a) Biological and (b) membrane tanks at Yunlin

5.2.3 Syndial, Porto Marghera, Italy

Syndial SpA is part of an international energy company engaged in oil and gas exploration as well as refining and petrochemicals production. The Porto Marghera plant in Venice produces various hydrocarbon products. The MBR installation at the site (Fig. 5-14) represents one of the most established large MBR installations for petroleum industrial effluent treatment, in this case from the ethylene/PVC process.

The original IWwTP was constructed by Lurghi in the 1970s. Legislation promulgated in 1998 required tighter limits of <120 mg/L COD, <10 mg/L TN and <2 mg/L N-NH_4 for discharges in the Venice region. This led Syndial to seek a technology capable of meeting the new stipulations. MBR technology was subsequently identified as the most appropriate for this duty, and ONDEO selected to provide the plant with GE as the membrane supplier. The plant began operation in 2005, following a successful six-month pilot trial.

The plant is based on a peak design flow of 1,980 m^3/h and an average daily flow of 38,400 m^3/d. The feedwater contains mean COD, BOD and TKN concentrations of 280, 155 and 70 mg/L respectively. There is both pre-denitrification, in accordance with the standard MLE process (Fig. 5-14), and post-denitrification, the latter due to a persistent recalcitrant organic nitrogen component.

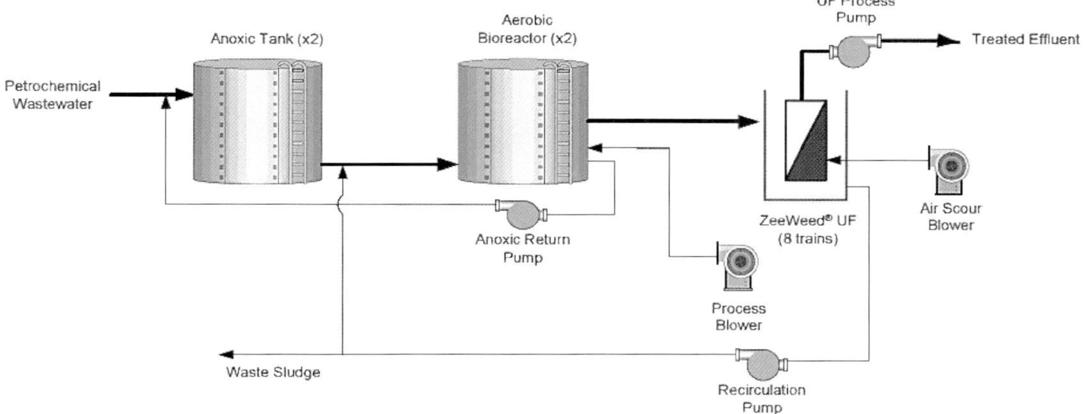

Figure 5-14 Porto Marghera MBR plant, schematic (GE, 2011)

The membrane component comprises 8 trains of 9 cassettes retrofitted into the existing aeration tanks. The cassettes are fitted with 44 GE *ZW500D™* modules (Section 4.2.4), each providing 31.6 m^2 membrane area and so yielding ~100,100 m^2 in total for the entire membrane separation. The flux is therefore 19.8 LMH at the quoted peak flow of 1,980 m^3/h, and 16 LMH at the average flow rate.

The plant has operated successfully since start-up, meeting the oxygen demand and N related standards as well as achieving significant removal of the toxic metal constituents.

5.2.4 Petrochemical plant, Sichuan, China

The petrochemical effluent treatment plant in Sichuan province treats 9,600 m^3/d of effluent and has been operational since May 2008. MBR technology was selected based on both treated water quality and footprint limitations, with the permeate used for irrigation and general purpose (such as road washing).

Wastewater at 15-30°C (25°C on average) containing 650 and 2,500 mg/L BOD and COD respectively is screened to 10 mm and then 1.5 mm (using a rotating drum screen) with intermediate grit removal. It is then equalised before passing on to the biological treatment process, which is configured for N & P removal. HRTs for biotreatment comprise 1.6 h anaerobic, 3.4 h anoxic and 6 h aerobic, with recirculation through the membrane tanks (2 h HRT) at a recycle ratio of 3Q. The MLSS concentration is 4,800 mg/L in the process tanks and 6,000 mg/L in the membrane tanks, maintained by an SRT of 15-30 d.

The two membrane tanks are each fitted with 20 cassettes (laid out as five rows of four units, Fig. 5-15a) containing 32 x 20 m^2 Motimo *FP* modules (Fig. 5-15b-c, Section 4.2.15), and thus a total membrane area of 25,600 m^2. This equates to a net flux of 16 LMH. The membranes are backflushed for 2 minutes at twice the instantaneous flux every 10 minutes. They are also air-scoured at 0.25 Nm^3/h per m^2 using two 45 kW blowers (hence 0.23 kWh per m^3 permeate for membrane aeration). This maintains the TMP at around 0.3 bar. Maintenance cleaning takes place weekly using a 1 h CIP with NaOCl followed by HCl, both at a concentration of 300-500 mg/L. This is supplemented with 4-6 h external recovery cleans using 1,000-5,000 mg/L NaOCl and 400-1,500 mg/L NaOH applied once or twice yearly as required.

The plant produces COD, BOD and ammonia levels of <50, <10 and <5 mg/L respectively.

(a) (b) (c)

Figure 5-15 (a) Membrane tanks, (b) membrane removal for external cleaning, and (c) membrane unit at Sichuan

5.2.5 Yanan Fengfuchuan oilfield reservoir, China

Three large-scale projects based on Pentair *X-Flow* UF membrane technology have been implemented in three Chinese oil fields (Daqing, Yanan Wangpiwan and Yanan Fengfuchuan) for produced water (PW) treatment and water recovery, all since 2012. Treatment of the UF filtrate by either RO or ion exchange provides low-salinity reinjection water, which improves overall oil production volumes. The two smallest plants at Yanan Wangpiwan and Yanan Fengfuchuan (Fig. 5-16) employ MBR technology.

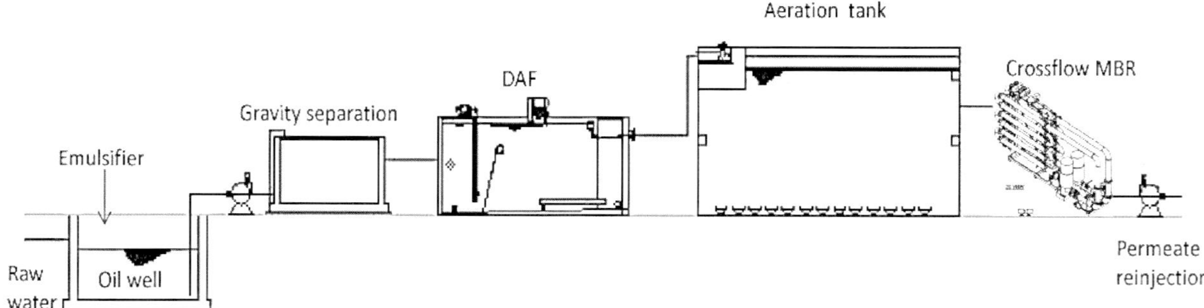

Figure 5-16 Schematic of effluent treatment plant, Yanan Fengfuchuan

The installation at Yanan Fengfuchuan has a capacity of 1,500 m^3/d. The membrane skid is based on 5 loops of 6 *Compact 27* modules (Section 4.3.1.4) per stream, thus offering a total area of 810 m^2 and a net flux of 77 LMH. The biotank is fed with PW pre-treated with a DAF and containing 20 mg/L TSS, and 45 mg/L of residual oil. The MLSS concentration is ~12,000-15,000 mg/L, with the sludge being recirculated at a flow of 200 m^3/h, or around 3Q, through the membranes at a feed pressure of 3.2 bar and an outlet pressure of 1 bar. The permeate water quality attained for reinjection into the reservoir is <1 mg/L TSS and < 1 mg/L oil.

5.2.6 Baosteel Shanghai, China

The wastewater treatment scheme (Fig. 5-17) at the Baosteel site in Shanghai, China, treats wastewater from a steel rolling industrial plant. The installation generates 3,800 m^3/d of wastewater containing emulsified oil and recalcitrant organic matter from acid washing, degreasing and galvanising operations. MBR technology was selected to provide the appropriate level of treatment and the plant was commissioned in May 2009.

(a)

(b)

Figure 5-17 The MBR plant at Baosteel: (a) the membrane tank, and (b) permeate lines

The wastewater (10,000-66,000 mg/L COD and 2,000-5,000 mg/L oil with a conductivity of around 4,000 μS/cm and a temperature of 5-40°C) is pre-treated by equalisation, and two-stages of pH adjustment and DAF prior to entering the biotreatment stage. The aerobic treatment employs a total HRT of 10 hours, the MLSS being held at 12,000 mg/L in the aerobic tank by an SRT of 30-90 days. Membrane separation is provided by 45 SINAP *150-200* cassettes (Section 4.1.20) each fitted with 200 x 1.5 m^2 panels. There are thus 8,960 panels in total offering a total area of 13,440 m^2 and a corresponding net flux of 12 LMH at the mean flow. The membranes are air-scoured at a rate of 0.48 $Nm^3/(m^2.h)$ and cleaned every 3 months with 5,000 mg/L hypochlorite and citric acid at a pH of 1-2.

The treated wastewater has a COD and oil concentration of ≤70 mg/L and ≤3 mg/L respectively. The water is partly used for on-site irrigation and the rest discharged.

5.2.7 Sinopec Luoyang, Henan Province

Sinopec Luoyang Company is a large refining and petrochemical company in the western Henan province of Central China with key businesses of oil refining, and chemicals and chemical fibre production. The chemical fibre operation generates over 1,200 kte/a of polymer and polymer fibre and an effluent stream of 4,800 m^3/d. The wastewater temperature varies between 20 and 40°C according to the production schedule, with feed COD, BOD_5 and ammonia levels of 200-400, 40-80 and 5-40 mg/L respectively. Since November 2004 this stream has been treated by an iMBR which is based on Memstar *SMM-1013* technology (12.5 m^2 "curtain" HF membrane modules, Section 4.2.11). The installation was undertaken by NOVO Envirotech (Guangzhou) Ltd as an EPC project (Engineering, procurement and construction).

Pre-treatment comprises equalisation combined with two stages of aeration and sedimentation to reduce the load onto the MBR (Fig. 5-18). The MBR process tank (Fig. 5-19) comprises a 1,200 m^3 anoxic zone and a 2,400 m^3 aerobic tank, providing an HRT of 18 h. There is no separate membrane tank, the MLSS increasing from 3,000-4,000 mg/L in the anoxic zone to

4,000-5,000 mg/L in the aerobic/membrane tank sustained by an SRT of 30-40 d. The feed is dosed with HCl at a rate of ~1 te/day to maintain the pH below 8 and suppress scaling. The wasted sludge is dosed with polyacrylamide at a concentration of 5 kg/te DS prior to dewatering on site.

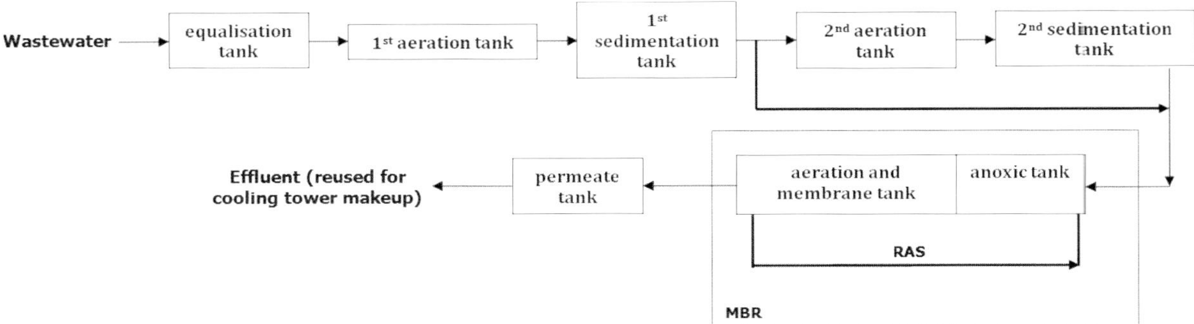

Figure 5-18 The Sinopec Luoyang plant, schematic

Figure 5-19 The MBR process tank at Sinopec Luoyang

Membrane separation is provided by nine trains each fitted with four cassettes of 48 modules (Fig. 5-19), yielding a total membrane area of 21,600 m^2. The membranes operate at a net flux of 9.3 LMH, based on 9 minutes filtration and one minute relaxation. This flux is sustained by weekly maintenance CIPs of 500 mg/L NaOCl and supplementary cleans with 1,000 mg/L NaOCl followed by 4,000 mg/L citric acid 2-3 times a year. Membrane air scouring is applied at a rate of 0.185 Nm^3/(m^2.h). The overall permeability is maintained between 50 and 140 LMH/bar.

Membrane separation for this application is challenged in particular by the high concentrations of cobalt and manganese in feedwater. These promoted scaling, producing a sharp decline in permeability and leading to filament embrittlement. The plant has otherwise performed satisfactorily, achieving BOD, COD and ammonia removals of 97%, 80% and 95% to leave residual concentrations of 2, 50 and 1 mg/L respectively. The overall *SED* is relatively low at 0.85 kWh/m^3, with the membrane air scour also contributing to the process air; these two components combined make up 85% of the MBR energy demand. The RAS pump, mixer, sludge discharge pump and sludge dewatering accounts for ~0.05 kWh/m^3 of the remaining energy demand whilst all other pumping operations (permeate, backwash and chemical dosing) demand ~0.08 kWh/m^3.

5.2.8 Intel, Qiryat Gat, Israel

The 9 MLD capacity (7.6 MLD average daily flow) plant at Qiryat Gat in Israel treats microprocessor fabrication effluent and has been in operation since Nov 2008. The effluent was originally treated for local agriculture irrigation, but a downstream RO unit was implemented in 2010 allowing the treated water to be reused for cooling tower make-up.

Wastewater at 27-30°C containing up to 400 mg/L COD and 80 mg/L N-NH_4/TKN is equalised for 3.2 hours prior to biotreatment. The MBR comprises two trains/streams including anoxic/aerobic basins of 742 and 1,000 m^3 respectively in each train, with transfer to four 85 m^3 membrane tanks at a recycle ratio of 5Q. The overall HRT is thus around 10 h.

According to available data to 2012 the membrane tanks each contained 3 x 48-module cassettes based on the GE *ZW500D™* module (31.6 m^2 membrane area, Section 4.2.4), thereby providing a net flux of 17.5 LMH. Scouring air was provided at a rate of 0.43 Nm^3/(m^2.h) and backflushing applied for 30s every 12 minutes. Maintenance cleaning with 250 mg/L hypochlorite is carried out twice a week, supplemented with weekly cleans with 2,000 mg/L citric acid.

The plant is operated at an MLSS concentration of 8.3 and 9.7 g/L in the biotank and membrane tank respectively, with a recorded MLVSS/MLSS ratio of 0.76-0.78 and an SRT in the region of 40 d. For data to 2012 COD removals were around ~96% (15 mg/L in effluent) and the TKN was removed down to levels of ~19 mg/L. The effluent TN concentration was attained by IPA injection into the anoxic tanks for C:N ratio correction, thereby meeting the required <25 mg/L limit imposed by the local EPA.

5.2.9 Bromide plant, Israel

The manufacturing plant in Israel produces bromo-organic compounds for flame retardant materials, generating an effluent which is designated "tough-to-treat" (or "TTT") by the MBR technology provider GE. The 2,500 m^3/d average daily flow plant (140 m^3/h peak flow) has been operating since March 2007. The treated effluent is discharged to evaporation ponds, though it is envisaged that ~40% of it could be reused following RO treatment.

Effluent at between 25 and 40°C (33°C on average) and containing 1,000 mg/L TOC and up to 1,000 mg/L ammonia is coarse-screened to 12 mm before undergoing physicochemical treatment, including neutralisation with caustic soda, and equalisation for ~24 h at the average flow. It then enters the aerobic tanks (3,000 m^3) and is recirculated through 4 x 60 m^3 membrane tanks, the HRT thus being ~31 h at the average flow rate. Dosing with antifoaming agent is occasionally required; the sludge is dewatered before being tankered off site.

The 4 membrane tanks are each fitted with 2 cassettes containing 48 x 31.6 m^2 *ZW500D™* modules (Section 4.2.4), and thus a total area of 12,134 m^2, generating a net flux of 8.6 LMH at the average flow. The filtration cycle is 6 mins, including 30 s of backflushing. The membrane operates at a TMP between 0.15 and 0.35 bar. There is a weekly maintenance CIP of 30 minutes based on 200 mg/L NaOCl and 1,000 mg/L citric acid. This is supplemented by quarterly 8 h-long recovery cleans with 1,000 mg/L NaOCl and 2,500 mg/L citric acid.

The plant provides effluent of <20 and <300 mg/L BOD and TOC respectively, with no nutrient removal required. It was still operating with the original membranes after six years, one year more than the warrantied life.

5.3 Pharmaceutical

5.3.1 Cosmetics effluent treatment, Madrid, Spain

The MBR plant at a PCP production facility located near Madrid treats 360 m^3/d of effluent. The effluent had previously been treated by an SBR downstream of a physicochemical treatment scheme comprising EQ, coagulation with PACl, neutralisation with NaOH, flocculation with a cationic flocculant and flotation. Changes to the pipework, cleaning systems and process units led to dramatic increases in the wastewater organic load and concentrations (Table 5-1) which deleteriously impacted on the performance of this plant. This, coupled with a new policy for water reuse and ongoing land restrictions, led to extensive trials of an MBR technology prior to its eventual implementation.

Table 5-1 Raw and pre-treated effluent characteristics, cosmetics effluent treatment, Spain

Parameter	*Raw*	*Pre-treated*
pH	7.1±0.1	
Conductivity (mS/cm)	0.9±0.1	1.6±0.8
COD (mg/L)	36,000±2,000	5,200±750
Soluble COD (mg/L)	23,600±1,300	4,500±950
TSS (mg/L)	8,000±700	1,300±300
VSS (mg/L)	5,600±500	900±200
Fat, oils and grease (mg/L)	1,700±300	234±86

The effluent is pre-treated by a 1.5 mm rotary fine screen before entering the aeration tank. The existing stainless steel SBR biotank has been reused for the MBR, with the aeration system upgraded to 1,000 Nm^3/h capacity so as to provide the required DO concentration (1-5 mg/L) at the higher MLSS concentration of 8-10 g/L (cf. 3.5 g/L for the SBR). The biological process generates 0.25 kg waste sludge solids per kg COD. Nutrient balancing by dosing with urea and K_3PO_4 is required to maintain the COD:N:P ratio at 100:5:1.

Sludge is transferred to the membrane tank at a recycle ratio of 4Q (Table 5-2). The membrane separation unit comprises two 500 m^2 *PURON® PSH500* (Koch Membrane Systems, Section 4.2.7) membrane cassettes immersed in a single 25 m^3 tank. The average permeate flux through the membranes is maintained at ~12 LMH (and the TMP at 12 LMH at 272 ± 97 mbar) using a filtration cycle of 6 min filtration followed by a 30 s backflush. Membranes are alternately intermittently aerated in cycles of 60 s at an SAD_m of 0.2 $Nm^3/(m^2.h)$ overall. The HF lumens are de-aerated for 120 s every 6 h and the air trap on the permeate line vented for 30 s every 20 cycles. A 90-minute weekly maintenance clean is employed using 1,000 mg/L NaOCl and 2 g/L citric acid, supplemented every 5 minutes-or-so.

Under steady-state conditions the treated effluent is characterised by a BOD_5 of 13 ± 5 mg/L, a COD of 98 ± 25 mg/L (corresponding to 98% removal), a conductivity of 2,600 ± 700 μS/cm, and a FOG concentration of 10 ± 5 mg/L. The plant allows intermittent operation during periods of low wastewater flow. A cost analysis indicates chemicals consumption accounts for 1.63 €/m^3, whereas the total energy cost for pre-treatment, biological reactor and sludge treatment is around 0.10 €/m^3. The cost of sludge treatment and disposal for both primary and secondary sludge is 0.84 €/m^3. The total operating cost of the whole treatment plant, based on a 5-year membrane life and a wastewater flow rate of 200 m^3/d for 240 d/y on average, is estimated at 3.3 €/m^3, 1.27 €/m^3 of this relating to the MBR and 1.19 €/m^3 to the pre-treatment.

5.3.2 Amgen, Puerto Rico

Amgen is a multibillion dollar international biotechnology company whose factory in Puerto Rico is subject to a zero liquid discharge (ZLD) restriction. The 1,900 m^3/d average flow (4,500 m^3/d peak flow) MBR plant installed at the site is upstream of an RO which recovers the effluent for reuse as boiler and cooling tower feedwater. The plant was designed by Ovivo and has been operating since March 2006.

Table 5-2 Membrane operating conditions, cosmetics effluent treatment, Spain

Parameter		*Values*	*Units*
Filtration flux		12	LMH
Filtration time		300 – 360	s
Backflush flow		30	LMH
Backflush time		20 – 30	s
Ventilation flow		6	LMH
Ventilation time		20 – 30	s
Ventilation frequency		20	cycles
Deaeration flow		25 – 30	m^3/h
Deaeration time		120	s
Deaeration frequency		4 – 6	h
Maintenance cleaning frequency		1	per week (168 h)
Maintenance cleaning flow		6	m^3/h
Dosing pumps			
Oxidant (NaOCl)	Concn.	1,000	mg/L
	Dosing	1,800	s
	Purging	600	s
	Air mix	300	s
Citric acid	Concn.	2,000	mg/L
	Dosing	1,800	s
	Purging	600	s
	Air mix	300	s
Recirculation IN	min	4Q	
Recirculation OUT	min	3Q	

The wastewater derives from pharmaceutical preparation and spent softener regenerant; it has a BOD and TKN of 1,670 and 186 mg/L on average and a temperature of 30-35°C. The water passes through a 2 mm drum screen before entering the bioreactor, comprising anoxic and aerobic zones of 500 and 1,580 m^3 respectively which, combined with the 490 m^3 membrane tanks, provide an HRT of 33 h at the average flow and 14 h at peak flow. The sludge is held at around 20,000 mg/L by an SRT of 61 d, and is recirculated through the membrane tanks at a rate of 4Q. Chemical dosing is applied in the form of alkalinity, for maintaining nitrification, and alum for P removal.

The four membrane tanks are each fitted with 4 Kubota *EK400* double-deck units (Section 4.1.9), 400 panels/unit, thereby providing 5,120 m^2 membrane area and thus a net flux of 8.9 LMH at the average flow and 21 LMH at peak flow. The membranes (warrantied for 10 years) are scoured at a rate of 0.53 $Nm^3/(m^2.h)$ and relaxed for 1 minute during each 10 minute filtration cycle to maintain the TMP between 30 and 100 mbar. Chemical cleaning using 5,000 mg/L NaOCl is applied 2-4 times a year, incurring 1-2 hours of downtime per clean, and this is supplemented with oxalic acid cleans at 10,000 mg/L once or twice yearly.

The plant has consistently produced mean BOD, TN and TP levels of 5, 35 and 3.3 mg/L respectively.

5.3.3 Sanofi Pasteur, Swiftwater, Pennsylvania, US

MBR technology was selected for the WwTP at the Sanofi Pasteur Inc. vaccine production facility when a rapid expansion in administrative, research and development, and production activities was planned, almost tripling the required treatment capacity to 3,200 m^3/d ADF (6,400 m^3/d PDF). MBR technology was selected as the most appropriate for this duty on the basis of footprint limitation and the requirement for water reuse. The plant was installed and commissioned on a fast track schedule, with commissioning completed in July 2008 – within one year of the project start - by the designers of the plant, Ovivo. The permeate is post-treated with UV before being substantially reused for seasonal landscape and spray irrigation, with less than half being discharged to a receiving surface water.

Feedwater at 15-25 (22°C on average) having a BOD and TKN of 276 and 38 mg/L respectively is passed through a 2 mm drum screen and on to the process tank. The MLSS is held at 10,500 and 12,500 in the process and membrane tanks respectively, the recirculation ratio being 5Q and the mean HRT and SRT being 10 h and 27 d respectively. The process tank comprises a 230 m^3 anoxic zone and a 740 m^3 aerobic zone to provide TN removal, with alum dosing employed for P removal.

The 290 m^3 membrane tank is fitted with 4 trains of 3 x *RW400* double-deck Kubota units (Section 4.1.9), each with 2 x 200 panels of 1.45 m^2 membrane surface area. The membranes are scoured at a rate of 0.14-0.41 $Nm^3/(m^2.h)$, 0.29 on average, and relaxed for 1 minute every 10 minutes. This maintains a net flux of 19 LMH and a TMP largely below 0.1 bar. Supplementary chemical cleaning is conducted 1-3 times a year with 5,000 mg/L NaOH and yearly with 10,000 mg/L citric acid.

The *SED* for membrane and process aeration, sludge transfer, permeate pumping and mixing is 0.29, 0.16, 0.18, 0.042 and 0.037 kWh/m^3 respectively, the total being 0.71 kWh/m^3. The plant consistently removes both the BOD and TKN down to below 5 mg/L.

5.3.4 Pharmaceutical plant, Taizhou, Zhejiang Province, China

A complete wastewater treatment plant for a multinational pharmaceutical plant in Taizhou, Zhejiang Province, China was completed in December, 2012. The plant was designed by the Portuguese company Valorsabio in collaboration with Shanghai MegaVision Membrane Engineering & Technology Co., Ltd. Implementation of the technology followed the successful completion of several months of on-site testing for system optimisation, this being demanded by the complex and challenging nature of the industrial effluent being treated. The wastewater flow was around 400 m^3/day, according 2010 data.

The retrofit solution makes use of the existing concrete tanks together with anaerobic pre-treatment with a UASB (Valorsabio, *UASB-PRO*), implemented in November 2011, and aerobic treatment based on Valorsabio's *Jet-Loop System*©®. Membrane separation is based on Shanghai MegaVision *FMBR-1.0-100* FS membrane panels (Section 4.1.14). These are arranged in 12 modules containing 100 x 1 m^2 panels (and thus a flux of around 14 LMH at the flow of 400 m^3/day) divided between two tanks. Further clarification by chemical coagulation/flocculation has been installed to remove non-biodegradable substances.

The *UASB-PRO* employs a new pulse process for the wastewater feed which addresses issues sometimes encountered in large UASBs relating to unbalanced feed distribution, short-circuiting inside the bioreactors and difficulties in stabilising the up-flow of wastewater within the design range. The anaerobic treatment reduces the COD loads by more than 85%, generating biogas which offsets the operating cost by its use for water heating.

The aerobic process operates at an HRT of less than 25% of the previously installed conventional activated sludge process. The COD is reduced from 1,300-1,400 to 600-800 mg/L.

5.3.5 Pharmaceutical plant, Spain

The pharmaceutical plant in Spain generates 350 m^3/day of effluent, which is treated using a Likuid Nanotek (Section 4.3.2) ceramic membrane-based sMBR prior to being discharged to sewer. The selection of MBR technology followed pilot trials which demonstrated it as being the most able to meet the required treated effluent COD concentration of <1,500 mg/L from an 18,000 mg/L COD feed. The plant was commissioned in 2002, initially with an FS polymeric membrane installed. However, the eventual presence of solvents in the effluent together with the relatively high temperature of the mixed liquor obliged the proprietor to look for a more robust technology when the plant needed to be upgraded to 350 m^3/day, since the membrane replacement frequency was considered too high.

The high loading rate of 4.5 kgCOD/(m^3d) and MLSS concentration of 15-20 g/L, maintained by an SRT of 23-30 d, sustains an *F:M* ratio of 0.3 kgCOD/(kgVSS.d) for this plant. At the design flow the 900 m^3 biotank provides an HRT of 62 hours. These conditions are then appropriate for the required >90% removal of COD from the effluent, its mean temperature generally being between 30 and 35°C.

The biological process employs a pure oxygen-based oxygenation system (In-Situ Oxygenation *I-SO™*, Praxair Inc.) to maintain high mass transfer efficiency at the elevated solids concentrations. Such systems have been demonstrated as being well suited to industrial MBRs where high MLSS concentrations and high organic loadings prevail, demanding high oxygen uptake rates (OUR) to sustain the process. OUR values for CAS wastewater treatment are typically in the range of 20-40 mg O_2/(L.hr), while in highly loaded MBR systems they can reach 50-150 mg O_2/(L.hr). Previous work also demonstrated the *I-SO™* system to provide oxygen transfer at very high solids concentrations because of the limited impact of solids on the alpha factor of the device. Since the use of pure oxygen requires lower gas flow rates it also reduces the stripping of VOCs (volatile organic carbons) and dissolved gases, as well as the propensity for foam formation.

There are 2 membranes skids, each comprising 12 vertically-aligned Likuid Nanotek *L37* modules with stainless steel pipework and fittings. The TMP ranges between 0.5 and 2.5 bar with a nominal mean value of 1.8 bar. The pressure and design crossflow of 3 m/s are sufficient to maintain an operating flux of 90 LMH. The net flux of 87 LMH is sustained by an operating cycle of 120 minutes of filtration with 30 s relaxation, along with the CIP. Chemical cleaning is applied eight times a year using a proprietary alkaline membrane cleaner (*Divos 120*, supplied by Diversey who are now part of the Amsterdam-based Sealed Air group), supplemented with an acid clean (*Divos 35*) three times a year. Both solutions are applied at a 1% v/v concentration.

Likuid Nanotek have at least 15 other industrial installations worldwide. In most cases the flux is between 90 and 105 LMH, with no backflush required to maintain the permeability. CIPs are generally applied monthly to quarterly; for example, the landfill leachate plant at A Coruña (Spain) employs a CIP every two months. The membrane has been shown to be very robust for similarly aggressive wastewaters (such as high temperatures, extreme pH levels and the presence of solvents or oxidizers) or highly-fouling wastewaters such as those laden with oil and grease.

5.3.6 Southern Taiwan Science Park (STSP) pilot plant, Chia Nan University, Taiwan

An extensive pilot trial based on full-scale MBR membrane modules has been conducted by the Department of Environmental Engineering and Science, Chia Nan University, at an industrial site in Taiwan. The work included measurement of the sludge and foulant characteristics.

The 10 m^3/d plant (Fig. 5-20a) comprised a 10 m^3 buffer tank followed by process and membrane tanks each having an 8.7 m^3 working volume (and thus providing 42 h HRT). The plant was challenged with wastewater at 16-28°C and generally between 800 and 12,000 mg/L COD concentration derived from combined pharmaceutical and septic tank effluent.

The MLSS concentration was increased from 6,000 to 12,000 over an initial operating period of 80 d, and was then operated at 12,000-17,000 mg/L solids concentration for a further 40 d. The viscosity was found to vary logarithmically with MLSS over the entire concentration range studied (Fig 5.20b). At a mean SRT of 40 days and COD loading of 0.099-6.844 kgCOD/(m^3.d), the *F:M* ratio was between 0.014 and 0.65 d^{-1} and the sludge production rate 0.035-0.072 kgSS/kgCOD. The sludge was recirculated through the membrane tank at a ratio of 5Q.

(a) (b)

Figure 5-20 (a) The pilot plant at STSP, and (b) the reported relationship between sludge viscosity and MLSS

The membrane tank was fitted with 2 x 12.5 m² *Flat Plat* iHF Motimo modules (Section 4.2.15), yielding a mean net flux of 16 LMH. The membranes were operated with two physical cleaning cycles: 30 seconds of relaxation every 4.5 minutes, and 6 minutes of backflushing every hour. Manual cleans (declogging) were conducted after 43 and 92 days when the TMP had risen to around 0.4 bar, though the TMP was generally between 0.1 and 0.25 bar for the majority of the operational period. A supplementary chemically-enhanced backflush was employed after 102 days.

During the 140 d period following acclimatisation (for 30 d) the recorded COD removal efficiencies were generally between 90 and 96%. Treatment costs, including amortised capital costs and depreciation, were evaluated as being between 1.27 and 1.75 €/m³. Whilst the technology provider specified a warrantied life of two years for this duty, the membranes were still operating after three years with only a few fibre breakages recorded over that period.

5.4 Pulp and paper

5.4.1 Gippsland Water Factory, Australia

The 35 MLD Gippsland Water Factory (GWF, Fig. 5-21a) generates 8 MLD of reclaimed water from domestic (16 MLD) and pulp and paper (19 MLD) influents. The scheme, designed by CH2MHILL, incorporates a number of unusual and ambitious features, including biological treatment of both the water and off gas streams, and has been operating since 2010.

The treatment scheme (Fig. 5-21b) includes preliminary treatment, anaerobic treatment, primary sedimentation, MBR treatment, biological nutrient removal, disinfection and RO, with an NF stage planned in the future. The scheme has segregated municipal and industrial wastewater streams, the WAS and industrial wastewater being treated in anaerobic lagoons (AL) prior to biological treatment via an MBR which then also oxidises the anaerobically-generated H_2S to elemental sulphur. Segregation of the higher salinity and more highly-coloured industrial effluent stream is critical in reducing the overall load on the MBR, and also in protecting the RO downstream of the municipal MBR component. The design also permits flexible use of the MBR membranes during peak municipal flows of up to 40 MLD, when all 12 of the cells can be harnessed to treat the flow rather than the 4 assigned to average daily flow conditions.

Each of the 12 cells are fitted with 7,220 m² of membrane area (Daigger et al., 2013), provided by 216 Evoqua *MemJet®* membrane modules, yielding a peak flux of 19 LMH at 40 MLD flow. With the normal configuration, employing 8 of the cells for treating 19 MLD of industrial effluent, the industrial MBR flux is around 14 LMH. Extensive pilot trials demonstrated that anaerobic pre-treatment reduced the fouling propensity of the downstream aerobic MBR membranes.

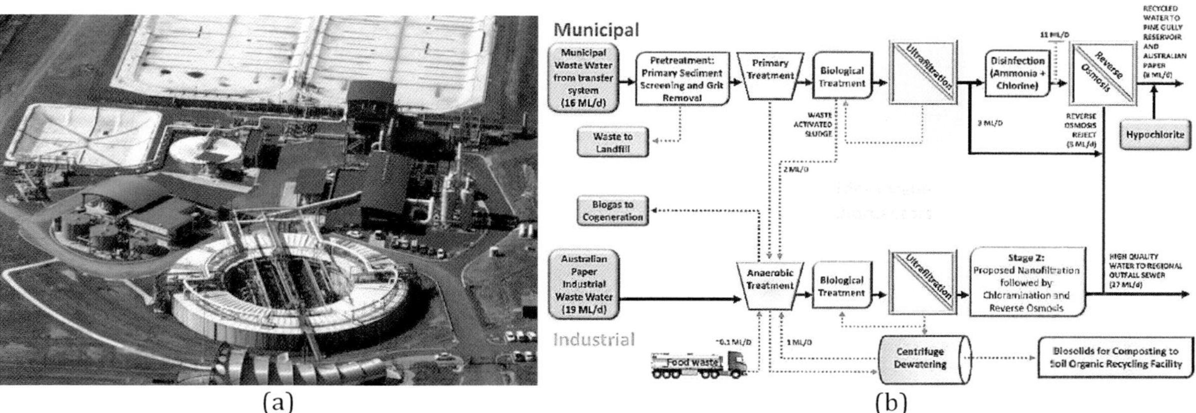

(a) (b)

Figure 5-21 Gippsland Water Factory (a) aerial view, (b) schematic (from Hodgkinson and Skeels, 2013)

The overall design of the industrial treatment component comprises (Daigger et al., 2013):

- Anaerobic reactors (2 off), 22,000 m^3 each. Lined earthen lagoon with membrane cover. Feed zone and effluent clarification zone with submerged launders.
- Industrial pre-filters (3 off), 25,230 m^3/day each. Automatic, self-cleaning units.
- Industrial bioreactors (2 off), 3,068 m^3 each. Aerated three-stage units of similar dimensions and configuration as domestic bioreactors.
- Industrial membrane tanks (8 off), 64 m^3 each. Evoqua/ *MEMCOR® MemJet®* units each containing 12 modules per tank with 7,220 m^2 of membrane area per tank.
- Dewatering solid bowl centrifuges (2 off) rated at 15 L/sec each.
- Cake hopper (1 off), 123 m^3 volume. Twin screw live bottom.
- Biogas desulphurisation (1 off), 9,000 m^3/day. Caustic scrubber with biological regeneration
- Cogeneration (1 off), 330 kW. Reciprocating engine.
- Odour control (4) 12,500 m^3/day each. Foul air scrubbed through bioreactors and then treated in two-stage biological scrubbers.

5.4.2 Pulp and paper plant, Arizona, US

The small (~20 m^3/d capacity) MBR plant (Fig. 5-22) at a pulp and paper pilot plant in Arizona was designed and installed by Layne Christensen Company to remove TSS and BOD to reduce the effluent surcharge incurred by its discharge to the sewer system. The MBR plant was fitted with downstream RO for pure water recovery. The pilot plant was operated between May and September of 2012.

Wastewater at 15-25°C (20°C on average), and containing an average 8,500 and 280 mg/L COD and TKN respectively, was pre-treated by primary sedimentation (8 h) prior to fine screening to 2 mm. Biological treatment was via a 2.3 m^3 anoxic zone and a 16.7 m^3 aerobic zone. The sludge was transferred at a recirculation ratio of 4Q to a 5.7 m^3 membrane tank fitted with a 4-module (24 m^2 total) Layne *Poreflon™* MBR cassette, based on the *SPMW-05B6* module (Section 4.2.17). The total HRT was thus around 30 h.

The plant was operated at an MLSS concentration of 10 and 12 g/L in the biotank and membrane tank respectively. A flux of 29 LMH was sustained by an air scour rate of 0.5-0.6 Nm^3/(m^2.h). Operation was between 0.07 and 0.4 bar TMP, with twice-monthly one hour-long maintenance cleans using 300 mg/L NaOCl and 500 mg/L NaOH. There was the option to supplement the CIP with nominally yearly or twice yearly recovery cleans by soaking for six hours in 3 g/L NaOCl and NaOH. These aggressive cleaning conditions are afforded by the chemical robustness of the PTFE membrane, which is warrantied for 7 years.

Figure 5-22 The Layne MBR plant for pulp and paper effluent treatment

The plant achieved COD, N-NH_4, TN and TP removals of 97%, 96%, 97% and 95% down to levels of 220 (due to high COD levels in the feed), 0.5-10, 19 and 4.1 mg/L respectively. For this small plant the permeate and sludge pumping operations contributed almost two thirds of the total *SED* of 1.61 kWh/m^3. The MBR system provided sufficient pre-treatment for the downstream RO system, and with appropriate chemical pre-treatment the RO membrane was successfully operated for approximately two months without membrane cleaning. The RO product water met or exceeded the paper-making water quality requirements.

5.4.3 Paper mill, Gomà-Camps SAU, Spain

The paper mill effluent treatment plant in the North East region of Spain was commissioned in July 2011. An MBR was selected primarily on the basis of the required treated water quality for discharge to the nearby river, the required effluent quality being <25 mg/L BOD, <100 mg/L COD from a feed composition of 2,500 mg/L COD, with nearly 3 mS/cm of conductivity. The design is based on a 720 m^3/d average flow, with daily and hourly peak flow of 960 m^3/d and 40 m^3/h respectively. The flow is met using 9 x *EK400* Kubota modules (Section 4.1.9) each providing 320 m^2 of membrane area, yielding average and peak daily fluxes of 10.4 and 13.9 LMH respectively.

Reactor aeration is at 4,040 Nm^3/h overall, provided continuously. The filtration cycle is 9 minutes filtration, 1 minute relaxation. The design MLSS is 12 g/L, maintained by an SRT of ~28 days and an HRT of 36 h at the mean daily flow rate. Nutrient dosing (urea and phosphoric acid) is provided to maintain the required C:N:P balance. The dosages are regulated with the residual concentration after filtration. The plant operates with a single tank for both process and membrane separation, with coarse bubble aeration for membrane scouring and fine bubble diffusion aeration (FBDA) for the process biology. The total energy demand at peak flow is 6.7 kWh/m^3, with 98% of this derived from aeration.

The membranes required ex-situ chemical treatment to decalcify them after around 2.5 years of operation, the regular CIP having been found to be insufficient to recover the permeability. Soaking in 3% of NaOCl for around 1 hour was found to be sufficient to completely restore the permeability. During this period the load to the effluent treatment plant was low enough to permit the ex-situ cleaning without overloading the remaining installed membranes.

Shortly after the ex-situ cleaning the plant experienced excessive loads due to process issues in the mill, leading to oxygenation problems in the effluent treatment plant which were compounded both by issues concerning the nutrient dosing and very large diurnal swings in the COD load. The decrease in DO to almost zero then led to a significant reduction in the filterability of the sludge, even when reduced to below 10 g/L, reducing the flux by a factor of three. Recovering the dissolved oxygen concentration to 0.5 mg/L and adjusting the N and P dosing did not improve the permeability significantly within a period of a week or so.

The challenge was subsequently met by reducing the MLSS concentration further to 8 g/L and allowing the DO concentration to be restored to its target level of 2 mg/L. Within 9 days the permeability had been restored to around 500 LMH/bar, and was subsequently maintained by caustic CIP (0.5%). The perturbation in the normal operation was attributed to abnormal contaminants, such as fibre and inks, possibly coupled with the nutrient dosing failure. The operators now have a procedure for dealing with future process upsets of this nature.

5.5 *Textile*

5.5.1 Bamberger Kaliko, Bamberg, Germany

A textile finishing company in Bamberg, North Bavaria had operated its own 400 m^3/d-capacity wastewater treatment plant for 20 years, the treated water either being discharged or reused for service water. Prior to 2008, the plant comprised a buffer/homogenisation tank, chemical treatment using ferric chloride and lime, a lamella separator, a conventional biological treatment stage, and sludge treatment. Discharged treated wastewater limits comprised COD <1,000 mg/L, pH 6-7 and < 30°C temperature.

Changes in production led to increased hydraulic and organic loads on the plant, such that it was no longer able to meet the required discharge standards. In addition, the surcharge for pollutant discharge was increased by the municipality, adding to the overall cost. This led the company to seek options for improved management of its effluent, given the challenges imposed by the discontinuous wastewater production and the wide variability in pH (6 to >13), temperature (25-80°C), waste sludge compostability, and pollutant concentration – the latter including intermittent intense colour.

A pilot MBR trial, designed and supplied by Huber and based on the company's *VRM®* technology (Section 4.1.6), was conducted using the clarified effluent as the influent (Huber, 2014). It became evident that upgrading the clarifier to a DAF was necessary to reduce the solids load to almost zero. The *VRM® 20/36* MBR pilot plant (Fig. 5-23a) comprised 108 m^2 membrane area, provided by 12 hexagonal plates, and was designed for an average permeate flow of 2 m^3/h which therefore equated to a net flux of ~19 LMH. Over a four-month period the MBR unit process was found to successfully remove 95%, 93% and 58% on average of the COD, N-NH_4 and P-PO_4 at mean feed concentrations of 2,810, 10 and 19 mg/L respectively. The outputs provided sufficient confidence to proceed to the full-scale plant.

The full-scale plant was installed downstream of the existing 250 m^3 equalisation tank. It comprised a 60 m^3/h capacity DAF, with preceding chemical precipitation and flocculation, a 250 m^3 aerobic tank and two *VRM® 20/300* membrane filtration modules each having 12.5 m^3/h flow capacity and 900 m^2 membrane area (hence ~14 LMH). The Huber *VRM®* is normally air scoured at a rate of between 0.15 and 0.25 $Nm^3/(h.m^2)$. The complete plant was commissioned in September 2008, with the DAF being installed first and the MBR subsequently.

(a)

(b)

Figure 5-23 (a) Pilot-scale and (b) full-scale Huber *VRM*® modules (Huber, 2014)

Initial challenges encountered at the plant were overcome through modification of the wastewater management. Acidification of the wastewater within the mixing and balancing tank was experienced during plant start-up, the cause being fractions of coating products within the wastewater containing starch. The pH was regulated by adding strongly alkaline production wastewater to counter this. Optimisation of the coagulation-flocculation process was conducted, with an aluminium-based coagulant combined with anionic polymer flocculant found to be both more effective and lower in operating cost than the previously employed set-up. Recirculation through the membrane tank was also adjusted, resulting in a higher quality effluent and reduced membrane fouling. Finally, it became apparent that some of the colourants used for production could not completely be removed, resulting in a coloured permeate which precluded its reuse. This was overcome by directing the 20 m^3/d of effluent affected to an additional holding tank for treatment by the physicochemical stages only, the treated wastewater subsequently being discharged to sewer rather than being reused.

Following the above optimisation the plant was shown to perform consistently, with COD values of up to 11,000 mg/L being reduced to below 200 mg/L permitting reuse of over 90% of the treated wastewater.

5.5.2 Jiangsu Shenghong Printing & Dyeing Co., Ltd, China

The 10,000 m^3/d MBR plant at Shenghong treats effluent from dyeing operations upstream of an RO plant, the treated water being reused for production processes. The plant was commissioned in October 2007, the EPC contractor being Xiamen Visbe Co. Ltd.

Effluent at 10-35°C (27°C on average) is equalised for 8 hours before passing on to the 5,000 m^3 biotank (50:50 anoxic:aerobic) where the MLSS is held at around 3,000 mg/L and recirculated to the 800 m^3 membrane tank. Sludge is wasted at 280 kgDS/day (i.e. 62 d SRT) and the HRT is 14 h at the mean flow.

The membrane tank holds 19,200 m^2 membrane area in six trains of four units, which each house 80 of the 10 m^2 United Envirotech/Memstar *SMM-1010* HF modules (Section 4.2.11). The net flux is maintained at 21 LMH and the TMP between 0.2 and 0.6 bar by backflushing at ~20 LMH for 1 minute in every 10 minutes of operation whilst air scouring at 0.2 $m^3/(m^2.h)$. Maintenance cleaning comprises a monthly soak for 1.5 hours in 200 mg/L NaOCl, with recovery cleans 1-2 times each year by soaking for 4 hours in 300-500 mg/L NaOCl and then in 0.5 wt% citric acid.

The plant achieves >90% COD removal down to effluent concentrations below 60 mg/L whilst incurring specific energy demand values in kWh/m^3 of 0.3 for both membrane and process aeration (and hence 0.6 kWh/m^3 for total aeration) and 0.006 kWh/m^3 for sludge transfer.

5.5.3 Textil-Service Klingelmeyer, Darmstadt, Germany

The laundering company Textil-Service Klingelmeyer in Darmstadt began seeking alternative options for water supply following a large increase in the cost of fresh water supply, and especially wastewater discharge, in the mid-1990s. The original CAS-based installation failed to achieve the required water quality for reuse, leading the company to seek alternatives. The iMBR ultimately installed at the facility followed pilot testing conducted under an EU *LIFE* programme: Laundry Innovative Waste-water Recycling Technology (LIWATEC, 2005; Hoinkis et al., 2012). The full-scale plant began operation early in 2007.

The MBR (Fig. 5-24) is designed to treat a flow of 200 m^3/d at a mean COD concentration of 1,120 mg/L and a TN of 160 mg/L. It is screened to 0.2 mm using a SWECO vibrating screen before passing on to a storage tank, demanded by the discontinuous nature of the laundering process. The water is then passed into the biological process tanks, offering 126 m^3 volume from two partially-buried concrete tanks of overall dimensions 7.0 x 4.0 x 5.5 m high – the top 1 m being freeboard. One of the tanks is fitted with 2 double-deck Kubota *EK300* cassettes (Section 4.1.9), each containing 600 x 0.8 m^2 panels and thus providing a total membrane area of 480 m^2.

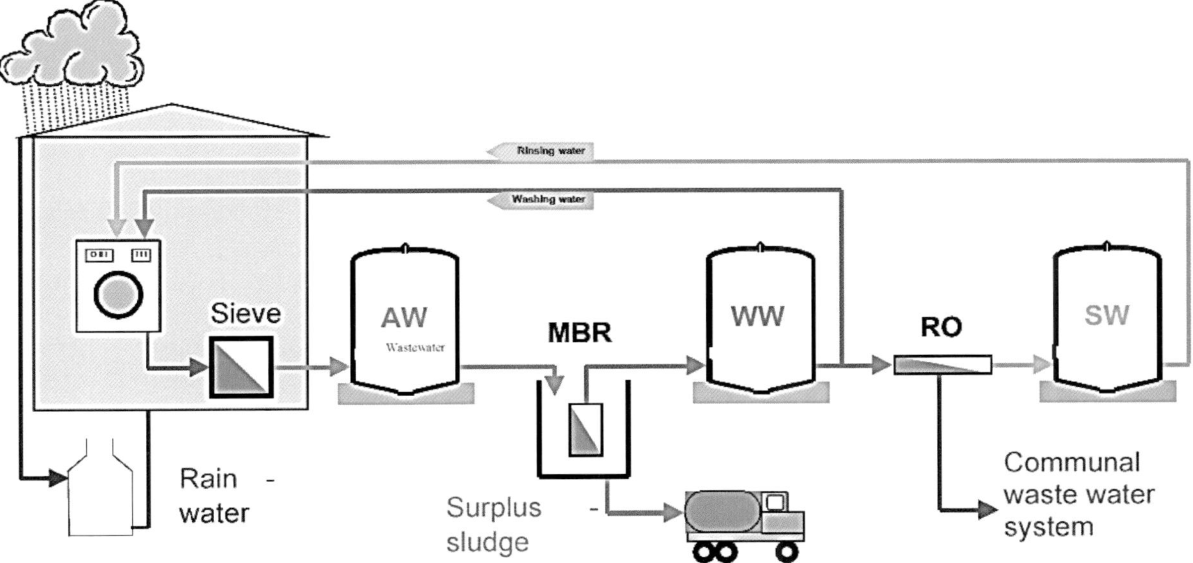

Figure 5-24 Schematic of the plant at Textil-Service Klingelmeyer (LIWATEC, 2005)

Air is delivered at a net rate of 430 Nm^3/h for the process, via FBD aeration, and 312 Nm^3/h for membrane air scouring. At the installed membrane area of 480 m^2 the SAD_m is thus 0.65 $Nm^3/(m^2.h)$ and the design net flux 17.4 LMH. In practice the net flux declined from 15 LMH to 12 LMH during the first year of operation, with the corresponding maximum permeability decreasing from 1,000 to 300 LMH/bar. No chemical cleaning was applied during this period, during which the MLSS increased from 3,000 to 10,000-15,000 mg/L and the feed COD from

~700 to over 1,000 mg/L. At the design SRT of 25 d and design HRT of 15 h waste sludge was generated at a rate of 0.13 kg DS per kg COD. The sludge holding tank is 60 m^3 in capacity, sufficient to hold 2-3 months of WAS.

The MBR permeate has a mean COD of 67 and a TN of 45 mg/L. Part of the MBR permeate is fed into an RO process, the permeate from which is blended with the MBR permeate at a ratio of between 1:1 and 1:2 (depending on the ion content) prior to being reused in the laundering operation. The RO treatment stage (based on Dow *LE400* low-pressure membranes) treats between 50 and 66% of the MBR permeate, operating at a flux of 25-30 LMH and a conversion of up to 80% at the feed pressure of 16 bar. The concentrate waste stream from this process, which makes up 12-15% of the total effluent from the laundry, is discharged to sewer.

5.5.4 Vanitec, Chennai, India

Vanitec is one of the 1,000-or-so tanneries operating in the State of Tamil Nadu in India. A 3,070 m^3/d capacity effluent treatment plant was installed at its factory in Chennai in 2011 to permit reuse of the treated water within the tanning process following further treatment by RO. The feedwater contains 3,000 mg/L COD, 1,200 mg/L BOD, 210 mg/L N-NH_4 and 0.9 mg/L P.

Effluent, generally between 25 and 30°C and flowing at an average rate of 2,500 m^3/d, undergoes 2 mm and 0.5 mm drum screening, along with chemical dosing for the EQ tank. It then passes to the biotank where the MLSS concentration is held at 15,000 mg/L. The sludge is recirculated through the Berghof *Bioflow* sidestream membrane skid (Section 4.3.3.2), operated at 5-6 bar pressure and 3-4 m/s CFV. The skid comprises 4 parallel loops of 6 x 53.4 m^2 modules (plus one "dummy" module) providing 1,282 m^2 membrane area in total. The net flux is thus 81 LMH on average, increasing to 100 LMH during peak flow periods. The membranes are chemically cleaned for ~2 hours monthly with acid and twice monthly with alkaline oxidant, both at 1 wt% concentration.

The process removes 90%, 97% and 99% of the COD, BOD and ammonia, leaving residual concentrations of 300, 12 and 6 mg/L of each respectively.

5.6 Landfill leachate

5.6.1 Songjiang District landfill site, China

The 150 m^3/d plant in the Songjiang District receives landfill leachate of around 14,000 mg/L COD and 1,800 mg/L ammonia at between 21-37°C, the BOD/COD ratio being ~0.23 on average, with the treated effluent being discharged to a common sewer. The MBR-based treatment scheme (Fig. 5-25) was installed in October 2004 by Shanghai JIFENG Environment & Equipment Co., Ltd., following the failure of the existing SBR-based treatment scheme to consistently achieve the required treated water quality.

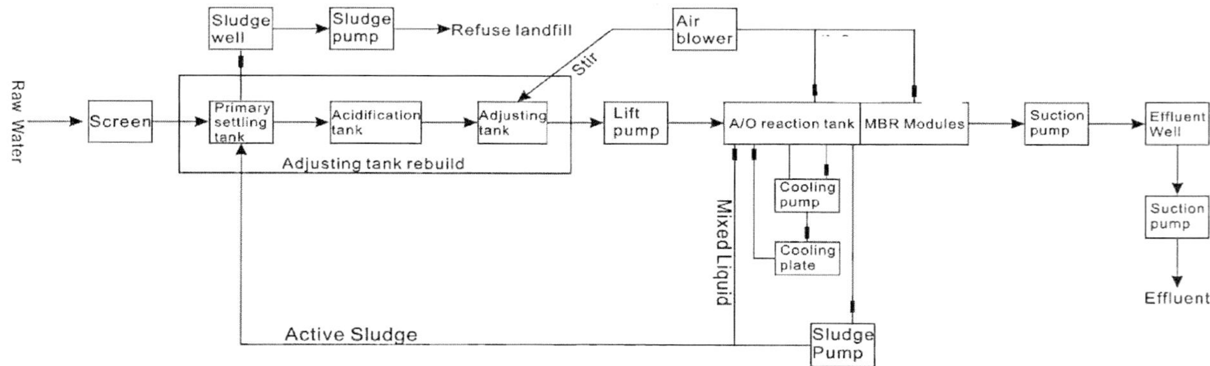

Figure 5-25 Treatment scheme at Songjiang

The feed is screened to 3 mm and pH adjusted before entering the biotank (Fig. 5-26a) where the MLSS is held at 11.2 g/L (13.3 g/L in the membrane tank). The total biotank volume is 2,030 m^3, encompassing a 1,350 m^3 anoxic and 547 m^3 aerobic/membrane tank. The membrane tank is fitted with 720 m^2 of membrane area provided from 8 x 60-panel double-deck SINAP *150* cassettes (Fig. 5-26b) at 1.5 m^2 per panel (Section 4.1.20). The membranes operate at a net flux of 8.7 LMH at the average flow, with each 5-minute filtration cycle including 60 s of relaxation. TMPs range between 0.1 and 0.4 bar and the SAD_m is 2.1 $Nm^3/(m^2.h)$. Recovery cleans every 20-60 days employ HCl to remove phosphate fouling. The MBR plant removes 97% of the feed COD, BOD and ammonia down to residual concentrations of <400, <150 and <60 mg/L respectively.

Figure 5-26(a) The process tank, and (b) the SINAP membranes at Songjiang

5.6.2 Ecopark De Wierde, Omrin, the Netherlands

The effluent treatment plant in Omrin, the Netherlands, is associated with an Ecopark and has been treated via an MBR plant since February 2003, prior to which the effluent had been tankered to a nearby sister site. Expansion of the site, and specifically the introduction of effluent from a new organic solid digestion plant, led to the decision to install a 20 m^3/h capacity MBR, comprising 15 m^3/h leachate and 5 m^3/h digestion waters. The treatment scheme (van Loggenburg et al., 2010) encompasses ~19 d of equalisation (9,000 m^3), the design being based on a wastewater composition of 3,000 mg/L COD and 1,000 mg/L TKN (80% NH_4). The discharge limit was set at 70 mg/L TN for the whole site, which at the time implied a 96% N-reduction by the MBR.

The biological system design is based on the Bardenpho process. It includes a 1,200 m^3 anoxic (denitrification, DN) zone with integrated methanol dosing based on NO_3 and redox measurements, a 1,200 m^3 aerobic (nitrification) zone and a 200 m^3 post-DN zone with methanol dosing as required. Membrane filtration originally comprised 2 lines of 10 m^3/h flow capacity provided by 1,360 m^2 iHF membrane area (hence a mean flux of 15 LMH), with a 4.5Q recirculation ratio and the membrane scouring air being reused for nitrification.

An increase in hydraulic and organic loading from the digestion wastewater (10-15 m^3/h of up to 15,000 mg/L COD and 4,500 mg/L TKN) demanded expansion of the plant to 35 m^3/h capacity. Whilst the original design could accommodate the increased organic loads with respect to biological treatment, the increased hydraulic load demanded a flux of up to 22 LMH which could not be attained by the existing membrane area. It was thus decided to retrofit a further 1,000 m² of area, provided by 2 Koch Membrane Systems *PSH500 PURON®* cassettes (Section 4.2.7, Fig. 5-27). The retrofit was completed in one day during September 2005, making use of the existing membrane tank infrastructure. The 74% increase in membrane area decreased the overall net flux back down to the original value of 15 LMH.

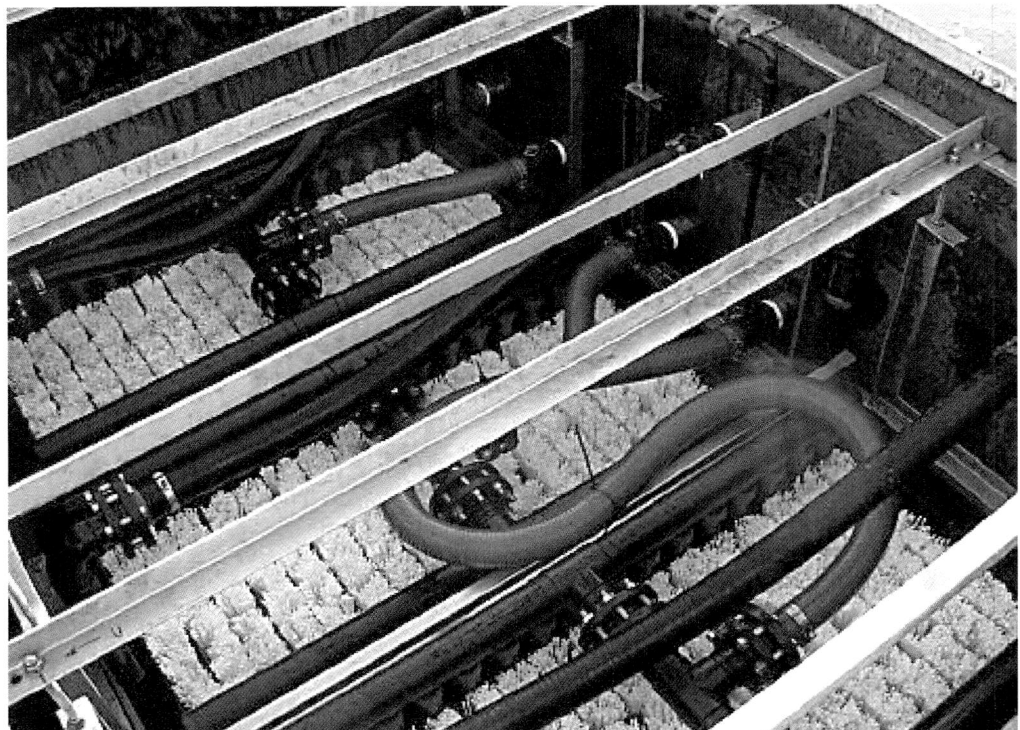

Figure 5-27 The retrofitted *PURON®* modules installed at Ecopark De Wierde

In mid-2006 the original membrane cassettes, which were experiencing problems of membrane permeability recovery and frame corrosion, were replaced with three *PSH500* cassettes. This increased the total membrane area to 2,500 m² across the two trains (2 cassettes in one train and 3 in the other), yielding a net flux of 14 LMH. As at 2007, the MBR plant provided ~85% removal of COD and >95% removal of TKN from mean feed concentrations of 7,100 and 1,300 mg/L respectively, the feed BOD/COD ratio being 0.28-0.42 (van Loggenburg et al., 2010).

5.6.3 Da Phuoc landfill site, Ho Chi Minh City, Vietnam

Da Phuoc is a landfill, operated by Vietnam Waste Solution, which takes waste from Ho Chi Minh City and has been in operation since 2007. The population equivalent is 7.4 million, with waste transfer volumes of 3,000 te/d and a total area of 30.6 hectares. The plant, designed for average and peak daily flows of 1,020 and 1,440 m³/d respectively, was selected on the basis of its small footprint and the requirement for downstream nanofiltration. It was designed by WEHRLE Umwelt GmbH (Section 4.3.3.4) and commissioned in February 2012. The effluent has average COD and ammonia concentrations of 2,800 and 780 mg/L respectively and the mean temperature is 25°C.

The effluent is equalised for around 12 hours and passed through a 0.4 mm slotted screen before entering the biotank, comprising 2 x 2,900 m³ aerobic tanks and a 1,000 m³ anoxic tank. The MLSS is held at around 15,000 mg/L by the HRT of 6.7 d and a design SRT of 29 d, though in practice the latter is much longer due to the low feed BOD of 330 mg/L. As of 2014 only one of the aerobic tanks was required to be operational due to the low loadings compared to the design values.

The sludge is recycled through the sidestream MT membranes at a ratio of 10Q and a CFV of 3.9 m/s. The skid comprises 4 stacks of 5 or 6 modules, offering a total membrane area of 594 m^2. At the applied TMP of 3 bar the operational flux achieved has been 93 LMH - somewhat higher than the design flux of 80 LMH. The permeability is sustained by automated cleaning with proprietary acidic and alkaline membrane cleaning reagents from WEHRLE.

The plant achieves COD removals of 67% down to 930 mg/L and ammonia removals of over 99%. The denitrification rate is 20% at an influent BOD/COD ratio of 0.1. The total plant *SED* (including the MBR, NF, sludge dewatering and cooling system) is 12.4 kWh/m^3 when the plant is running at full capacity, the NF contributing ~1.9 kWh/m^3.

5.6.4 Riederberg landfill site, Austria

The landfill leachate treatment plant at Riederberg replaced an existing failing RO plant. The MBR plant was selected based on its small footprint, and has downstream nanofiltration for biorefractory organics removal. It was commissioned in January 2012. The plant is challenged with leachate having mean COD and ammonia concentrations of 15,000 and 3,000 mg/L respectively, the temperature being 20-28°C. The plant has a design capacity of 65 m^3/d (100 m^3/d peak) with respect to the UF stage but has been operated at flows as high as 140 m^3/d.

The plant, designed by WEHRLE Umwelt GmbH (Section 4.3.3.4), is fitted with a 750 m^3 buffer tank and a 0.8 mm bag filter inlet screen, as well as a 1 mm basket screen for the RAS. Caustic soda is dosed to maintain a neutral pH for nitrification. The biological treatment comprises a 60 m^3 anoxic tank and a 550 m^3 aerobic tank, thereby offering an HRT of 9.4 days under average flow conditions. The MLSS concentration is ~15,000 mg/L. The membrane skid is fitted with VSD pumps allowing TMP and CFV ranges of <1.8-3 bar and <2-3.8 m/s respectively. The lower limits correspond to low energy (LE) operation appropriate for low flows, whereas the upper limits relate to high flows and fluxes. This allows flexible, intermittent operation of the membrane separation system according to the level in the aerobic tank.

The membrane skid contains 4 x 27.2 m^2 modules providing 108.8 m^2 total membrane area with an option to increase to 136 m^2. The membranes are chemically cleaned monthly with acid and alkaline solutions, supplemented with recovery cleans with proprietary WEHRLE chemicals as and when required.

The MBR stage achieves around 60% COD removal and >99.5% ammonia removal upstream of the NF, with the final NF permeate COD being <300 mg/L.

5.6.5 Hazardous waste landfill site, Michigan, US

The MBR technology, provided by Dynatec, was selected for a landfill site in Michigan on the basis of it being the best available technology (BAT) for providing the highest treated water quality possible based on advanced biotreatment. The plant, commissioned in 2009, treats an average flow of 250 m^3/d of hazardous landfill leachate at a temperature of 18-24°C and with COD and ammonia levels of 6,100 and 400 mg/L respectively.

The leachate is screened to 1 mm and chemically treated for heavy metals removal using a proprietary reagent. The 150 m^3 aerobic tank operates at an HRT of 15 h and an SRT of 23 d, maintaining the MLSS at around 20,000 mg/L at the load received. No denitrification is required at this site, and the flow requires no equalisation.

Membrane separation is provided by 7 x 27 m^2 MT membrane modules in a single loop. Sludge is pumped at 4 m/s CFV through the sidestream, providing a recirculation ratio of 15Q and a flux of 93 LMH – 57% higher than the design flux. The membranes are chemically cleaned twice-weekly by flushing with 300 mg/L NaOCl followed by 3 wt% citric acid.

The plant reduces the COD levels to below 1,100 mg/L (>82% removal) and the ammonia levels to below 10 mg/L (>99.75% removal). The plant has run well, with the net flux exceeding the design value despite the higher than anticipated MLSS concentrations due to periodic lapses in sludge management. Additional sludge dewatering capacity has been installed and clean frequency increased to weekly during those periods of elevated MLSS concentration.

5.7 Ships

5.7.1 Ship effluent treatment, RWO Marine Water Technology

The installation, which features on board a number of *Celebrity Cruises* cruise liners, is in response to regulations relating to black water discharge, as specified in MEPC 159(55) under MARPOL Annex IV (Section 3.8.1). RWO's *MEMROD®* system is installed on five of the ships owned by Celebrity Cruises (operated by Royal Caribbean Cruises Ltd) operating in the Mediterranean and Caribbean seas throughout the year. All five ships, which represent one third of Royal Caribbean's cruising fleet, have passed the required compliance test.

On average there are between 4,500 and 4,800 people on these vessels, and the system on board a single vessel has been designed for a maximum flow of 2,460 m^3/d. All black, grey, galley and food waste reject wastewater is treated in a single system, providing a BOD load equivalent to a town of 25,000 population. Wastewater generated onboard (Table 5-3) is sent to the MBR along with wastewater from the centrifugal decanter (which treats the excess sludge) and the scrubber (which treats the exhaust fumes).

Table 5-3 Ship wastewater stream characteristics

Wastewater stream	*Volume (m^3/day)*	*BOD (mg/L)*
Accommodation water	873	200
Laundry grey water	276	300
Galley grey water	272	2,000
Black water (vacuum system)	91	3,000
Food waste reject water	20	20,000-50,000
Scrubber water	18	17,000 (pH 3-4)
Total	**1,550**	

All streams pass through the screening system and are then sent to the small (10 m^3) collecting chamber (Fig. 5-28). The streams have been programmed to offset each other; for example, the grey water and black water are split 70% to 30% to balance the organic loading to the system. The water is then sent to one of the two bioreactors where air is pumped into each tank fed by three blowers (two duty and one standby) rated at 1,300 Nm^3/h at 0.5 bar. Air scouring to the membrane tanks (Fig. 5-29d) is provided via two more (duty/standby) blowers rated at 2,100 Nm^3/h at 0.5 bar. The tanks are all fitted with an exhaust fan taking the excess air up to the smoke stack and keeping the tanks at a slight underpressure.

The optimum bioreactor MLSS is 9-11 g/L with an average HRT of 8.5 h. The feed pump transports the sludge to the submerged membrane tank (*MEMROD®*) where the MLSS is kept between 11-16 g/L to maintain the efficiency of the membranes. The membrane separation comprises 72 modules housed in six units providing just under 5,000 m^2 of total membrane surface and yielding a mean flux of around 11 LMH. A CIP process, fully automated and requiring no more than 30 minutes of manual intervention a week for minor adjustment, is designed to clean each membrane module monthly or as needed according to the maximum allowable TMP. The sludge temperature varies from 34 to 40° C in the Caribbean to a little less during the summer months in the Mediterranean. Excess sludge is automatically drawn off, amounting to about 40 m^3/day.

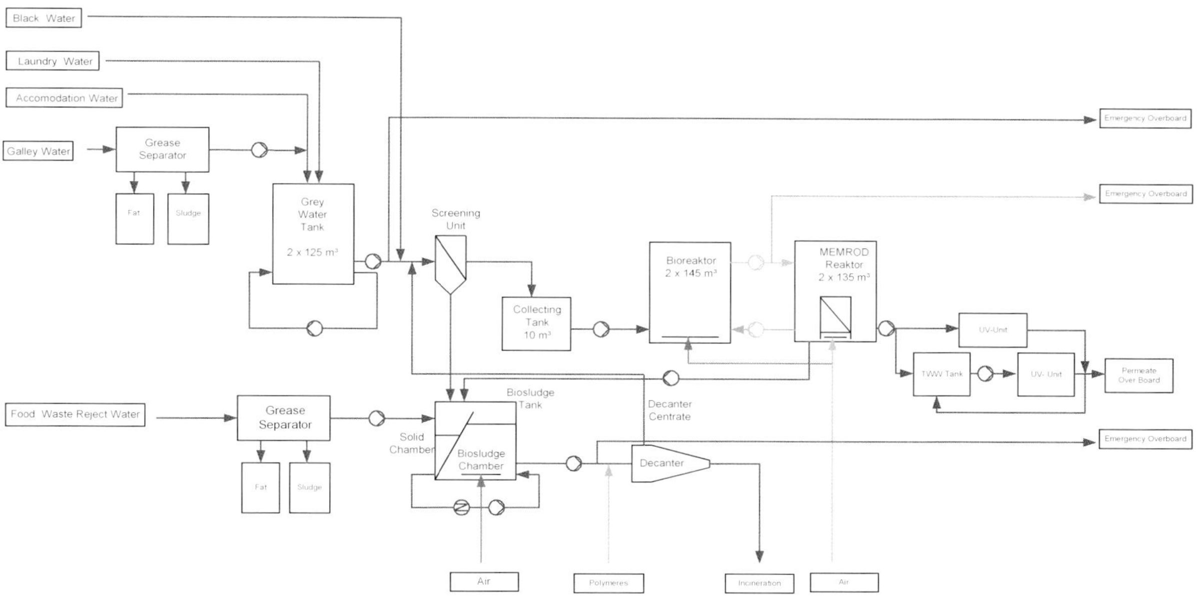

Figure 5-28 Plant schematic of ship effluent treatment, RWO *MEMROD®* system

(a) (b)

(c) (d)

Figure 5-29 RWO *MEMROD®* system: (a) recirculation pumps with a dosing station (foreground) with the grey water pumps in the background; (b) permeate flow lines for the six membrane units; (c) permeate pump and (d) membrane tank

In addition to the municipal wastewater from the accommodation, laundering, the galley, and black water, it has also been necessary to integrate the decanter water, comprising ~40 m^3/day, food waste reject water, 20 m^3/day and scrubber water, which adds a further 17-22 m^3/day (Table 5-3). The scrubber water is directed to the grey water storage tanks, where the low pH is easily neutralised. The other two streams are integrated into the collection chamber. Despite the heterogeneous nature of this wastewater, it is readily treated by the MBR technology. Moreover, the system is able to accommodate the new N and P limits for discharged water since aeration can be adjusted to generate an intermittent anoxic zone to attain higher total N removal. The dosing stations (Fig. 5-29a), already set up for anti-foaming chemical dosing, can be used to introduce polyaluminium chloride.

The operation of the plant has not been without some challenges. The collection tank was found to be too small to allow rigorous mixing of all the streams, the available space constraining its size. This has been addressed through redirecting some of the streams: whilst the scrubber water is fed into the grey water tank, the decanter stream is fed directly to the bioreactor. It also became evident that the membrane modules needed to be readily accessible in the event of a required change of membranes. All parts inside the tanks requiring maintenance must be accessible from the outside so that the sludge does not need to be emptied for basic routine maintenance and service (for example the declogging of the foam/water sprinklers). Moreover, pre-treatment using a screen rated at less than 2 mm is recommended to prevent fibres from reaching the membranes and causing ragging. Finally, it was evident that the exhaust fan propellers (located on Level 14) needed to be more robust to prevent them from wearing out, since this causes a slight overpressure in the tanks and makes maintenance more challenging.

Despite the challenges, the *MEMROD®* technology has been viewed favourably by the client. The systems continue to produce effluent consistently passing the compliance tests with relatively low O&M costs.

5.7.2 Ship effluent treatment, Wartsila

Wartsila Water Systems Ltd (previously known as Hamworthy Water Systems Ltd) is a dedicated marine wastewater treatment systems supplier. The company developed sidestream crossflow sMBR technology in the early 2000s, in response to the very stringent Alaska Regulations for cruise ships black and grey discharges. Similar discharge limits have also been required by the USEPA Vessel General Permit (VGP) for the grey water discharges from all ships inside US waters (Section 3.8.1). As of 2014, the Wartsila system has been installed and operated on >35 cruise ships, and the company has the largest market share of the MBR technologies in this sector.

The wastewater typically passes through a 2 mm screen and is then equalised for 8 hours and fine-screened before being fed into the biological tank. Effective grease separators are essential where galley grey water treatment is required. The sMBR can operate at MLSS concentrations of 15,000 to 20,000 mg/L, resulting in a smaller reactor volume. Depending on the need for nitrification, the *F:M* ratio can be designed from 0.1 to 0.3 d^{-1}, with a typical sludge yield being in the range 0.6-1.0 kgVSS/kgCOD.

The mixed liquor is filtered through 4-10 banks of 6-8 Pentair *Compact 27* UF membrane modules (Section 4.3.1.4), the precise number required for active service depending on the flux, at a feed pressure of up to 4 bar and a CFV of around 2-3 m/s. The flux ranges from 40 to 100 LMH, according to the characteristics of the wastewater, and in particular according to whether the grey water is being treated separately. The membranes are chemically cleaned every 4-6 weeks by a simple cleaning-in-place procedure, using small quantities of weak acid followed by a proprietary enzyme solution. The permeate product has a COD of <100 mg/L and total faecal coliforms of < 1 MPN per 100 mL.

5.7.3 Ship effluent treatment, ROCHEM/ULTURA

ULTURA provides a sidestream MBR technology (*Bio-FILT®*, Section 4.3.1.5) for effluent treatment and water reuse on board ships. The MBR permeate from the 503 m^3/d plant installed on a cruise ship in 2012 is treated by UV for general purpose use or, in the case of high-purity reuse duties such as boiler feed, a reverse osmosis stage can be added to provide demineralised water. The plant (Fig. 5-30) was originally installed with nitrification only, the intention being to upgrade the process technology to meet the MEPC.227(64) standard for N and P removal (Section 3.8.1).

The feedwater, which ranges from 20 to 40°C in temperature and has an average COD of up to 2,700 mg/L, is the combined black water and grey water from the ship toilets and kitchens/washrooms respectively. The MBR is designed to produce effluent having concentration limits of 25 mg/L BOD and 125 mg/L COD.

Figure 5-30 The membrane skids (on left) with biological process tanks (on right), *Bio-FILT®* system

The effluent passes through coarse and fine self-cleaning vibrational sieves, the finest being 0.15 mm rating, along with a minimum of 6 hours of flow balancing. The MLSS concentration in the bioreactor is kept at around 15,000 mg/L. The sludge is passed through a skid comprising 7 loops, each with 15 membrane modules, each module being fitted with 10 x 0.68 m^2 cassettes of FS membranes. The total area offered is thus 714 m^2, and the corresponding net flux is 29 LMH, although it may be higher if membrane operation is intermittent. The flux is maintained by a feed pressure of 4 bar and a crossflow of less than 1.5 m/s, and the membranes are air-backflushed every 30 minutes. Chemical cleaning with 2.5% sodium hydroxide and EDTA is scheduled 10 times annually.

5.8 Mixed industrial/municipal wastewater

5.8.1 Dalsung, South Korea

The 25,000 m^3/d PDF effluent treatment plant at Dalsung, an industrial park in South Korea, receives ~70% of its feed from food processing, steel production, textile dyeing and paper manufacturing plants, the remaining 30% being municipal wastewater. It was retrofitted to a conventional works comprising 6 mm coarse screening, degritting, 2.8 h primary sedimentation and 4.5 h flow equalisation provided as pre-treatment. The fitting of the MBR, with 1.5 mm fine screening downstream of the degritter, obviated the secondary clarification and provides 15,000 m^3/d of water for direct reuse by two of the paper production companies, the remaining water being discharged. The MBR, an example of the Econity *KSMBR* technology (Section 4.2.2), has been operating since June 2009. It was installed by SsangYong Engineering and Construction Co., Ltd. under the auspices of the consultant Hanjo Engineering Co., Ltd. for the client Environmental Management Corporation (EMC).

The plant provides biological nutrient removal, thus employing supplementary alum dosing at high P loads, with a 0.6-1.0 recycle ratio to the 1,728 m^3 anaerobic zone. There are two "dynamic state" bioreactors of 2,773 m^3 total volume combined with a 583 m^3 anoxic zone and a 2,830 m^3 membrane tank, providing a 7.6 h HRT at peak flow. The dynamic state bioreactors are aerated intermittently with, at any one time, one reactor operating continuously under anoxic conditions and the other operating in batch mode with a sequence of anoxic, oxic and anoxic conditions to provide simultaneous nitrification and denitrification. The anoxic zone reduces the RAS DO to <0.2 mg/L before entering the anaerobic tank. An SRT of ~44 d maintains the MLSS at 8,000 mg/L in the membrane tank.

Membrane separation is based on the Econity *KMS 6007CF* stack, containing 112 cartridges (7 cartridges high, 8 long and 2 deep) of 18 m^2. This provides a membrane area of 88,704 m^2 in total, and so a net flux of 12 LMH at peak flow. Filtration is conducted on a 12 minutes on/3 minutes relaxation basis with air scouring at a rate of 0.12-0.16 $Nm^3/(h.m^2)$, 0.145 on average, which maintains a TMP of 0.065-0.53 bar. Maintenance cleaning takes place every two weeks using a 2 h CIP using 2,000 mg/L NaOCl. Recovery cleans with 4,000 mg/L NaOCl are then applied as required.

The plant operates at a total energy demand of ~0.7 kWh/m^3, with 0.343, 0.027 and 0.086 kWh/m^3 for aeration, sludge pumping and permeation respectively. The plant generally achieves a permeate water quality of <10 mg/L for both COD and TN and <1 mg/L TP.

5.8.2 Taixin Binjiang WwTW, China

The 40,000 m^3/d capacity Taixin Binjiang wastewater treatment plant in Jiangsu province has been treating effluent from an industrial estate (10,000 m^3/d) combined with a domestic wastewater stream (30,000 m^3/d) since July 2010. The primary contractor for the work was NOVO Envirotech (Guangzhou) Ltd., with Taixing Wastewater Treatment Company Ltd. The feed includes textile and dye wastewater, pesticides, pharmaceutical intermediates, fine chemicals and other mixed wastewater partially of industrial origin. It is complex in composition, with a high level of colour (from the dyewaste), TDS and recalcitrant organic components, as well as fluctuating temperature. The treated water has to meet GB 18918-2002, the Chinese standard for pollutants in treated municipal wastewater which was implemented in 2003, to permit its discharge to a surface water.

The wastewater, whose temperature varies between 15 and 40°C, contains 300, 600, 50, and 3.3 mg/L of BOD, COD, N-ammonia and P respectively, as well as 1,000 Hz units. The domestic wastewater undergoes conventional coarse screening (25 mm) and grit removal whilst the industrial stream receives 8.5 hours of equalisation. Both streams are separately screened to 5 mm, with the domestic receiving a supplementary 1 mm screen.

The 10,000 m^3/d industrial effluent stream undergoes primary clarification in a 1,000 m^3 clarifier with a 200 m^3 flocculation tank, followed by PACl (500 kg/d, dosed at 6.3 mg/L as Al). The combined streams then pass on to a BNR process comprising anaerobic, anoxic, aerobic and membrane tanks of 1,333, 2,667, 8,000 and 4,000 m^3 respectively, providing an HRT of 9.6 h on average. At the recycle ratio of 3Q, the sludge concentration in the process and membrane tanks is 5,000-6,000 and 7,000-8,000 mg/L respectively, the SRT being around 40 d. The process tank is dosed with PACl for chemical P removal at a rate of 1 te/d (7 mg/L as Al_2O_3). Sludge dewatering on site consumes a further 30 kg/d polyacrylamide (5 kg/te DS).

Membrane separation comprises 8 trains of 20 stacks fitted with 48 United Envirotech/ Memstar *SMM-1520* membrane modules (Section 4.2.11), each providing 20 m^2 of membrane area. They are air-scoured at a rate of 0.11 $Nm^3/(m^2.h)$ which sustains a net design flux of 11 LMH and a permeability generally declining from 145 and 43 LMH/bar between maintenance cleans. The membranes are relaxed for one minute within an 8-minute cycle and chemically cleaned every two months with 1,000 mg/L NaOCl followed by 2,000-4,000 mg/L citric acid.

BOD, COD, N-ammonia and P are respectively removed down to 2, 50, 1 and 0.4 mg/L, and colour down to 30 Hz units. The total recorded *SED* for the plant is very low at 0.6 kWh/m^3, two-thirds of this figure deriving from membrane and process aeration and the remainder from all pumping and dewatering operations. The membrane warrantied life is five years and has already lasted for four of these.

5.8.3 Tongjiang WwTW, Heilongjiang province, China

The 10,000 m^3/d plant at Tongjiang in Heilongjiang province treats a 70:30 mixture of municipal and food industrial effluent, and was commissioned in July 2010. It was selected following successful pilot trials, on the basis of footprint limitation combined with the requirement for reuse of the treated water which demands a permeate turbidity of <0.5 NTU. The permeate is treated by RO, chlorination, ozonation, or UV according to the end reuse application locally.

Pre-treatment comprises a 10 mm bar coarse screen and a 1.5 mm drum fine screen with intermediate grit removal. Biological treatment encompasses anaerobic treatment (1.5 h) and anoxic treatment (2 h) for biological nutrient removal, along with 8 h of aerobic treatment at the mean flow rate. The sludge is recirculated through the membrane tank at 4Q, yielding a residence time of 2 h. The MLSS concentration is 5,600 mg/L in the biotank and 7,000 mg/L in the membrane tank, maintained by an SRT of 10-30 d.

The membrane separation component comprises two tanks each containing 6 units of 96 x 20 m^2 Motimo *FP* modules (Section 4.2.15) in double-deck configuration (Fig. 5-31a), providing 23,040 m^2 of membrane area overall. The membranes operate at an instantaneous flux of 22.6 LMH and a mean flux of 18 LMH at the 10,000 m^3/d design flow, the membranes being backflushed for 2 minutes at twice the instantaneous flux during each 10-minute filtration cycle. The membranes are scoured at a SAD_m of 0.25 $Nm^3/(m^2.h)$ using two 30 kW blowers (and hence 0.144 kW per m^3 permeate for membrane aeration, cf. 0.168 kW/m^3 for process aeration). The total energy demand is around 0.5 kWh/m^3.

Weekly maintenance cleaning for one hour with NaOCl and HCl, both at a concentration of 300-500 mg/L, is supplemented with recovery cleaning with 1-5 g/L NaOCl and 0.4-1.5 g/L NaOH every 6-12 months. The plant produces permeate having a COD and ammonia of below 50 and 5 mg/L respectively from a feed of 450 mg/L COD and 50 mg/L TN.

(a)

(b)

Figure 5-31 The membrane tanks at Tongjiang WwTW during (a) installation and (b) commissioning

5.9 Appraisal

A review of available data provided for the 51 case studies in Section 5.1-5.8, along with supplementary data from ten case studies taken from The MBR Book (Judd, 2011) reveals a few general trends in treatability and selection of design parameters overall.

5.9.1 Aerobic

5.9.1.1 Biological process parameters

It is generally considered that the BOD/COD ratio provides an indication of treatability, and there are evidently some effluents which are significantly more biorefractory (i.e. resistant to biodegradation) than others such that COD removal is limited even at extended HRTs. Thus for food and beverage effluents, for which the reported BOD/COD ratios are predominantly above 0.5, MBR treatment achieves more than 90% COD removal in all cases and at least 95% on average for 75% of the installations reported (Fig. 5-32a). For landfill leachate, on the other hand, for which reported mean BOD/COD ratios are mainly below 0.3, only a third of the plants listed achieve 90% COD removal. The trend across the different wastewaters is similar for residual COD levels (Fig. 5-32b). Trends for all other industrial wastewater types appear to lie between these two extremes. Effective ammonia or TKN removal is almost always achieved, invariably down to below 20 mg/L regardless of application provided nitrification is not inhibited. In two thirds of the cases where ammonia and/or TKN levels were reported removal was down to <5 mg/L. This is perhaps unsurprising given the generally long SRTs afforded by MBR technology, which tends to encourage the development of the slow-growing nitrifiers.

The reported performance in general, in terms of COD and N-NH_4/TKN removal, appears to be somewhat better than that indicated by the predominantly bench-scale data reviewed by Lin et al. (2012). This is evident from a comparison of COD removal data for the most extensively reported application of food effluent treatment, for which eight sets of data are provided by Lin et al., compared with the 19 case studies from Section 5.1 (Fig. 5-32a). Clearly, much larger data sets are needed for a more rigorous comparison but it is possible that this disparity reflects the acclimatisation period, since full-scale plants must necessarily achieve steady-state operation.

Patterns in MLSS concentration are less apparent. For certain immersed technologies and/or applications MLSS concentrations as low as 3,500 mg/L – comparable with a CAS – have been employed in conjunction with low recirculation ratios (<3Q). This then promotes concentration polarisation in the membrane tank, with sludge concentrations as much as double those in the process tank. This operational mode is perhaps motivated by the improved aeration efficiency in the process tank at the lower MLSS concentrations (Fig. 2.7), as well as slightly reduced sludge pumping costs. For the majority of the iMBR cases presented, however, the recirculation ratio is 4-5Q, yielding no more than a 25% increase in solids concentration in the membrane tank cf. the process tank.

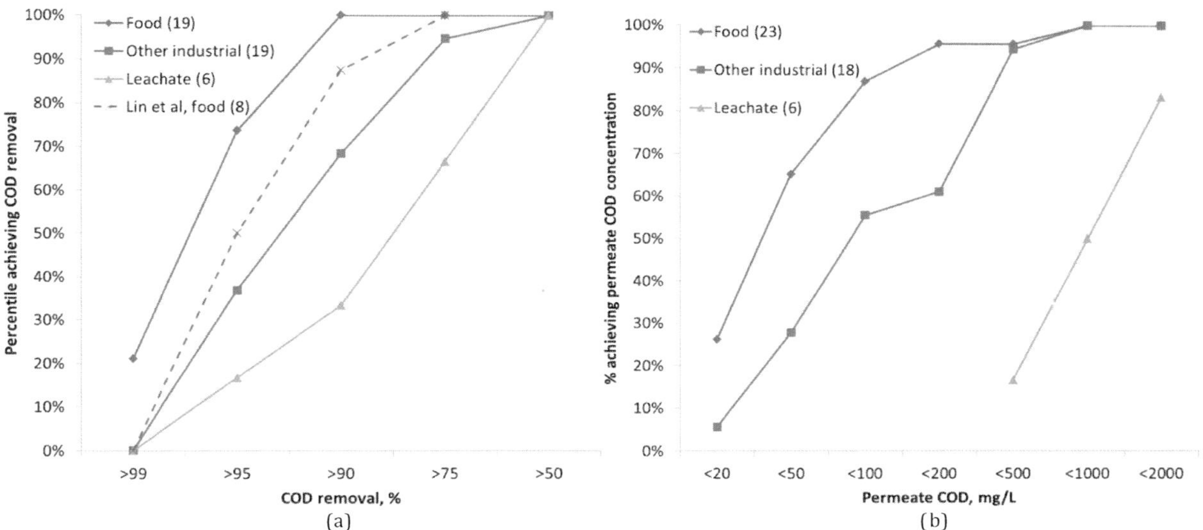

Figure 5-32 Overall trends in COD removal (a) % removed (dashed refers to data by Lin et al., 2012), and (b) permeate COD concentration, for 52 and 47 reference sites respectively.

5.9.1.2 Physicochemical process parameters

Membrane air scour rates for iMBRs do not appear to differ from those employed for municipal wastewater treatment for the same technology. Similarly, crossflows and pressures applied to A-L sMBRs appear to be the same as those used for municipal wastewater treatment. The physical and chemical cleaning parameters also do not generally change with application: relaxation and backflush cycles tend to be technology-specific, along with CIP protocols. However, the diversity in industrial effluent composition means that the risk of fouling by key inorganic foulant species, such as calcium carbonate scale, metal hydroxides and phosphates, may be higher for industrial applications and demand bespoke chemical cleaning measures.

Design fluxes selected for industrial effluents are, on the other hand, more conservative than for municipal applications. Values vary significantly with wastewater type and, as would be expected, technology type. Flux data are very highly scattered: there is no apparent trend with application for the 13 food and beverage data or the 23 industrial installations data (all data other than leachate) for which data are available. Indeed, both the mean and standard deviation values for these two groups of data are very similar (14.4 vs. 14.1 LMH mean flux; 37% vs. 34% standard deviation, Table 5-4). However, an examination of the range of fluxes for the different wastewater types (Fig. 5-33) reveals that 34 of the 39 iMBRs reviewed operate at a mean flux between 5 and 30 LMH.

The data also reflect the challenging nature of landfill leachate treatment. For the four examples of pumped sMBRs provided the fluxes attained are considerably lower than those reported for food effluent treatment using the same MBR process configuration (81 vs. 142 LMH on average, Table 5-4). Similarly, for the immersed configuration, the two flux data provided for iMBRs used for leachate treatment are both below 15 LMH (Fig. 5-33).

Table 5-4 Summary of flux data, 48 reference sites

Sector(s)	*Technology*	*Min*	*Max*	*Ave*	*SD*	*No. data*
Food	iMBR, FS & HF	8.6	25	14.4	37%	13
	sMBR, pumped	113	185	142	20%	5
	AL-sMBR	31	44	39	18%	3
Other industrial	iMBR, FS & HF	8.5	28	14.1	34%	23
Leachate	sMBR, pumped	54	93	81	23%	4

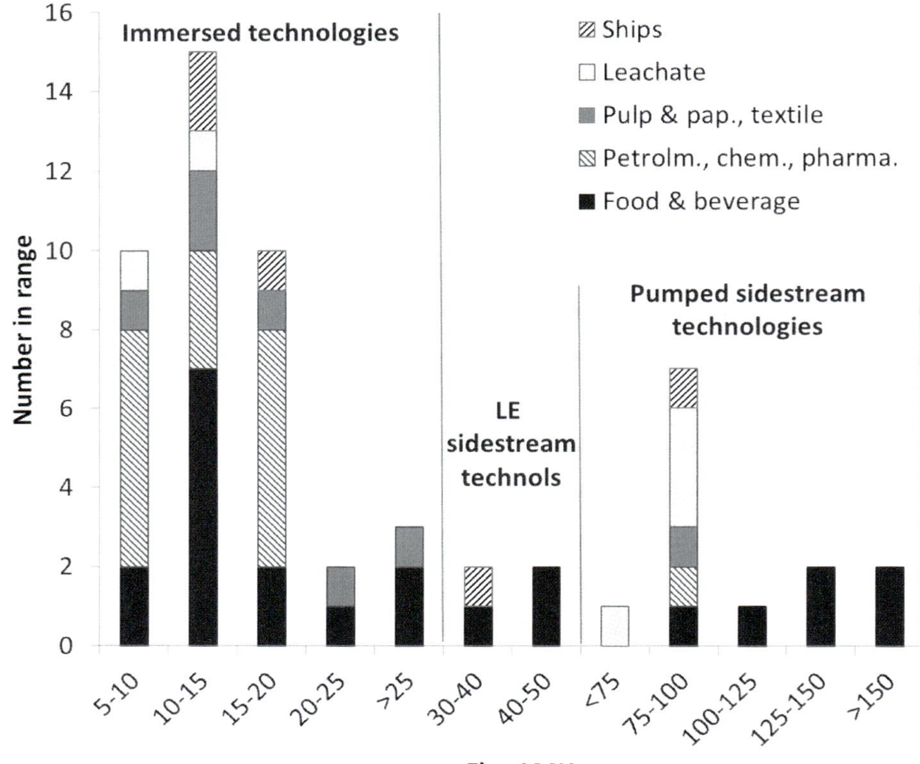

Figure 5-33 Summary of case study flux data by application and technology (57 reference sites): immersed, low-energy sidestream (including air-lift) and conventional pumped sidestream

5.9.2 Anaerobic

For feed COD concentrations between 12,000 and 57,000 mg/L, all of the anaerobic MBR technologies achieved removals of >99% and residual COD levels of <500 mg/L when challenged with high loads of readily biodegradable organic carbon (predominantly food and beverage industry effluents). As with any anaerobic treatment process, no nutrient removal is achieved and thus the viability of the process depends upon the required treated water quality in this regard and/or downstream removal of the N and P.

Reported fluxes are generally in the range of 4-6 LMH for iFS systems and 15-25 LMH for sMT systems, reflecting the highly fouling nature of the anaerobic biomass. There are some pilot studies (Robles et al., 2014) which suggest that optimised backflushing and CIPs combined with relatively low mixed liquor concentrations can allow fluxes as high as 20 LMH to be attained.

5.9.3 General

The relatively small flows usually involved mean that energy demand is not necessarily a major consideration for industrial effluent treatment. Moreover, the high organic loads generally imposed by most such effluents mean that process air delivery for aerobic treatment makes up a large proportion of the energy demand. This being the case, improving the energy efficiency of the membrane filtration component may provide comparatively little benefit. For small flows, a far greater proportion of the specific cost (cost per unit treated water volume) is the labour cost. It is generally more cost effective, over any reasonable amortisation period, to adopt a conservative approach to design and operation to reduce manual intervention as far as possible.

A key component of any industrial effluent treatment process is flow equalisation. This may be constrained by available space, though for retrofit projects aerobic and/or secondary clarifier tanks made redundant by the fitting of MBR technologies can be converted to EQ to mitigate against the effects of shock loads. Almost all the case studies provided in Sections 5.1-5.8 include at least eight hours of EQ. Differences in quality of effluents discharged from different operations within an industrial process suggest that segregated treatment of the different streams should provide efficiency gains, for example for refinery or textile processing wastewaters, though effluent stream segregation is rarely carried out in practice.

Reviews of MBR process technology invariably conclude that the process offers key advantages over competing conventional processes, most obviously the higher water quality provided and reduced footprint incurred. For industrial wastewater reuse the two-stage MBR-RO process is becoming the automatic preferred option. However, these advantages should not be allowed to deflect from the significantly greater complexity, both in design and operation, of the process compared with a CAS or SBR plant and the accompanying increased risk of process upset. For small plants typical of industrial effluent treatment, where the cost of unscheduled intervention may be very onerous, a conservative design to minimise maintenance requirements is more critical than for the larger municipal plants, particularly where specialist knowledge is unavailable on site. This inevitably increases the cost, making the viability of MBR technology most critically dependent on the amortisation (or payback) period selected.

References

Daigger, G.T., Hodgkinson, A., Aquilina, S., and Burrowes, P. (2013). Creation of a sustainable water resource through reclamation of municipal and industrial wastewater in the Gippsland Water Factory. *J. Water Reuse Desalination* **03.1** 1-15.

GE (2011). Syndial SpA Porto Marghera, GE Power and Water CS-SYNDIAL-INDWW-EN-1106-NA, May 2011.

Grant, S., Christian, S., Vite, E., Juarez, V. (2010). Anaerobic membrane bioreactor (AnMBR) pilot-scale treatment of stillage from tequila production. In: Proc. *IWA 12th World Congress on Anaerobic Digestion*, Guadalajara, Mexico.

Hodgkinson, A., and Skeels, P. (2013). Gippsland Water Factory – membranes at work. *Water* Feb 2013 edition 1-5.

Hoinkis, J., Deowan, S.A., Panten, V., Figoli, A., Huang, R.R. and Drioli, E. (2012). Membrane bioreactor (MBR) technology – a promising approach for industrial water reuse. *Procedia Engng.* **33** 234–241.

Huber (2014). Huber Technology, waste water solutions. Membrane technology for wastewater recycling in textile industry. http://www.huber.de/huber-report/ablage-berichte/industry/membrane-technology-for-wastewater-recycling-in-textile-industry.html (accessed July 2014).

Judd, S., and Judd, C. (2011). The MBR Book, 2nd Edition, Butterworth-Heinemann (Oxford, UK).

Lin H., Gao, W., Meng F., Liao, B.Q, Leung, K.T., Zhao, L., Chen, J. and Hong, H. (2012). Membrane Bioreactors for Industrial Wastewater Treatment: A Critical Review. *Crit. Revs. Environ. Sci. Technol.* **42** 677–740.

LIWATEC (2005), Laymen Report, Textil-Service Klingelmeyer GmbH & Co. KG.

Robles, A., Ruano, M.V., Ribes, J., Seco, A., Ferrer, J., and Judd, S. (2014). Immersed anaerobic MBRs: are they viable? http://www.thembrsite.com/features/immersed-anaerobic-mbrs-viable/ (accessed July 2014).

Singh, K.A., Burke, D., and Grant, S. (2010). Anaerobic flat sheet membrane bioreactor treating food processing wastewater: pilot-scale performance. In: Proceedings of IWA 12th World Congress on Anaerobic Digestion, Guadalajara, Mexico.

van Loggenburg, T., Herold, D., Kullmann, C. (2010). Membrane bioreactors (MBR) for landfill leachate treatment, International Biennial Conference & Exhibition, Durban, 18-22 April.

Annex 1 MBR operational energy and cost calculations

A Process biology

The base parameter values provided for the wastewater for treatment by an activated sludge process technology are as follows:

Biological and oxygenation parameters

COD removed	ΔS_{COD}	mg/L	1,000
TKN removed	ΔS_{TKN}	mg/L	50
COD content of biomass	λ_{COD}	kgCOD/kgMLSS	1.1
TKN content of biomass	λ_{TKN}	kgTKN/kgMLSS	0.095
Actual sludge yield	Y_{obs}	kgMLSS/kgCOD	0.35
Density of air	ρ_A	g air/m^3	1,230
Concentration of oxygen in air	C'_A	g O_2/g air	0.21
Standard oxygen transfer efficiency per unit depth	$SOTE$	m^{-1}	0.045
Depth of aerator in tank	y	m	5
Blower energy equation constant (based on SI units)	k	-	6.5
Blower inlet pressure	$P_{A,in}$	Pa	106,000
Blower outlet pressure ~ $P_{A,in} + P_y$	$P_{A,out}$	Pa	156,000
Mass transfer correction factor, solids	α	-	0.5
Mass transfer correction factor, salinity	β	-	0.95
Mass transfer correction factor, temperature	γ	-	1.00

The operational expenditure (OPEX) can be determined from the equations for:

a) Oxygen demand from COD and TKN loading in g/m^3 O_2, Equation (8):

$$D_{O2} = \Delta S_{COD}(1 - \lambda_{COD}Y_{obs} - 1.71\lambda_{TKN}Y_{obs}) + 1.71\Delta S_{TKN} - 2.86\Delta S_{Nitrate}$$

b) Aeration demand from calculated oxygen demand in Nm^3 air/m^3, Equation (9):

$$Q_A/Q_F = SAD_{bio} = D_{O2}/(\rho_A C'_A \, SOTE \, y \, \alpha \, \beta \, \gamma)$$

c) The specific aeration energy demand for air pumping in Ws/Nm^3 Equation (10):

$$E'_A = k\,P_{A,in}\left(\left(\frac{P_{A,out}}{P_{A,in}}\right)^{0.283} - 1\right)$$

The specific energy demand for biological process aeration ($E_{L,bio}$) in units of kWh/m^3 treated water is then the product of these two parameters:

$$E_{A,bio} = E'_A \, SAD_{bio}$$

Inserting the relevant values into Equation (8):

$$D_{O2} = 1{,}000\,(1 - 1.1\times0.35 - 1.71\times0.095\times0.35) + 1.71\times50 - 2.86\times8 = 621 \text{ g/m}^3$$

From Equation (9):

$$SAD_{bio} = 621/(1{,}230\times0.21\times0.045\times5\times0.5\times0.95\times1.00) = 22.5 \text{ Nm}^3/\text{m}^3$$

From Equation (10):

$$E'_{A,bio} = 6.5\times106{,}000\,((156/106)^{0.283}-1) = 80{,}000 \text{ Ws/Nm}^3$$

So $E_{A,bio} = 22.5\times80{,}000 = 1.8\times10^6 \text{ Ws/m}^3 = 1.8\times10^6/(1{,}000\times3{,}600) = 0.5 \text{ kWh/m}^3$

B Immersed membrane

The base data provided for the cost determination are listed below (see also Table 2-2). $E'_{A,m}$ is assumed to be the same as $E_{A,bio}$, which is justifiable if the aerator is at the same depth. $E_{A,bio}$ in units of kWh/m^3 is 80,000/(3,600 x 1,000) = 0.0222.

Membrane operation and cost parameters

Flux (assumed 18 LMH)	J	m/h	0.018
Specific aeration demand against membrane area	SAD_m	Nm^3/(h.m^2)	0.25
Specific energy demand for aeration ~ $E'_{A,bio}$	$E'_{A,m}$	kWh/m^3	0.0222
Specific energy demand for permeation	$E_{L,m}$	kWh/m^3	0.015
Membrane life (assumed 8 y)	t	h	70,000
Energy demand of sludge pumping (power/sludge flow)	$E_{L,sludge}$	kWh/m^3	0.02
Recycle ratio: membrane-bio tank + between biotanks	R	-	4
Electrical energy cost	L_E	\$/kWh	0.12
Membrane cost	L_M	\$/m^2	80
Chemicals cost, 10-15% hypo and 50% citric	L_C	\$/m^3	0.008

The OPEX is then given by Equation (13):

$$\begin{aligned}\text{OPEX} &= L_E(E'_{A,m}SAD_m/J + E_{L,sludge}R + E_{L,m} + E_{A,bio}) + L_M/(J\,t) + L_C + L_W + L_L \\ &= 0.12\times(0.022\times0.25/0.018 + 0.02\times4 + 0.015 + 0.50) + 80/(0.018\times70{,}000) + 0.008 \\ &= 0.12\ \text{USD/m}^3\end{aligned}$$

C Pumped sidestream membrane

In the case of a pumped sidestream technology fitted with horizontal modules there is no air scouring of the membrane, the shear being created by the crossflow. The energy demand is therefore determined from Bernoulli's equation (Equation 12):

$$E_m = [H + \Delta P/(\rho g) + v^2/(2g)]\,\rho g/(\varepsilon_{tot}\,\theta)$$

where appropriate values may be:

H = static head = 2 m
ΔP = applied pressure = 350,000 Pa (i.e. 3.5 bar)
g = 9.81 m/s^2
ρ = 1,000 kg/m^3
v = 3.5 m/s
ε_{tot} = 55% (the product of all efficiency values associated with the pumping)
θ = conversion.

The conversion refers to the proportion of the retentate passing through the permeate channels which is converted into permeate. This in turn can be determined from the membrane module characteristics (specifically the membrane area and the available cross-sectional area) and the net flux. Appropriate values for a food and beverage application might be:

Membrane area, A = 33 m^2
Flux, J = 150 LMH
Permeate flow = 4.95 m^3/h
Tube diameter = 0.2 m

So, the total cross-sectional area of the tube is:

A_{xsa} = 3.141 × 0.2^2/4 = 0.0314 m^2

If only 85% of this cross section corresponds to the membrane channels then the available cross-sectional area is

$A_{xsa,\,av}$ = 0.85 × 0.03141 = 0.0267 m^2

The retentate flow is the product of the crossflow velocity v and the above available cross-sectional area:

$$Q_R = A_{xsa,\,av}\, v = 0.0267 \times 3.5 \times 3{,}600 = 336 \text{ m}^3/\text{h}$$

The permeate flow per module is the product of the flux and the module membrane area

$$Q_P = J\,A = 150 \times 33 \;/\; 10{,}000 = 4.95 \text{ m}^3/\text{h}$$

If the sludge passes through eight such tubes then the total conversion is:

$$\theta = 8 \text{ x } 4.95 \;/\; 336 = 0.118, \text{ or } \sim 12\%$$

So,

$$SED_m = [2 + 350{,}000/(1{,}000 \times 9.81) + 3.5^2/(2 \times 9.81)] \times 1{,}000 \times 9.81/(0.55 \times 0.12)$$
$$= 1.6 \text{ kWh/m}^3$$

This figure ($E_{L,m}$ for the sidestream MBR) displaces the terms for energy demand relating to membrane air scour, permeate pumping and sludge transfer in Equation (13):

$$\text{OPEX} = L_E(E_{L,m} + E_{A,bio}) + L_M/(J\,t) + L_C + L_W + L_L$$
$$= 0.12 \times (1.6 + 0.50) + 80/(0.018 \times 70{,}000) + 0.008$$
$$= 0.32 \text{ USD/m}^3$$

Annex 2 Common conversions

Common conversion factors between Imperial and metric membrane technology parameters are provided below.

Imperial units	*Metric units*	*Imperial to metric*	*Metric to Imperial*
°F	°C	Subtract 32, multiply by 0.55	Multiply by 1.8, add 32
ft	m	0.305	3.28
GFD	LMH	1.70	0.588
GFD	m/d	0.0408	24.5
GFD/psi	LMH/bar	24.6	0.0406
hp	kW	1.34	0.746
inches	m	25.4	0.0394
lbs	kg	0.454	2.20
MGD	MLD	3.79	0.264
psi	kPa, bar	6.89, 0.0689	0.145, 14.5
SCFM	Nm^3/h	1.70	0.588
$SCFM/ft^2$	$Nm/(m^2.h)$	18.3	0.0546
sq ft	sq m	0.0929	10.8
U.S. gal	litres	3.79	0.264

SCFM Standard cubic feet per minute
psi Pounds per square inch
MGD Megagallons per day (U.S.)
MLD Megalitres per day
GFD Gallons per square foot per day
LMH Litres per square metre per hour

INDEX